AF906046

Design Science and Innovation

Series Editor

Amaresh Chakrabarti, Centre for Product Design and Manufacturing, Indian Institute of Science, Bengaluru, India

The book series is intended to provide a platform for disseminating knowledge in all areas of design science and innovation, and is intended for all stakeholders in design and innovation, e.g. educators, researchers, practitioners, policy makers and students of design and innovation. With leading international experts as members of its editorial board, the series aims to disseminate knowledge that combines academic rigour and practical relevance in this area of crucial importance to the society.

Gavin Brett Melles · Christian Wölfel
Editors

Design for a Sustainable Circular Economy

Research and Practice Consequences

Springer

Editors
Gavin Brett Melles
Faculty of Health, Arts and Design
Swinburne University of Technology
Hawthorne, VIC, Australia

Christian Wölfel
Industrial Design Engineering
TU Dresden
Dresden, Germany

ISSN 2509-5986 ISSN 2509-5994 (electronic)
Design Science and Innovation
ISBN 978-981-99-7531-0 ISBN 978-981-99-7532-7 (eBook)
https://doi.org/10.1007/978-981-99-7532-7

© The Editor(s) (if applicable) and The Author(s), under exclusive license to Springer Nature Singapore Pte Ltd. 2024

This work is subject to copyright. All rights are solely and exclusively licensed by the Publisher, whether the whole or part of the material is concerned, specifically the rights of translation, reprinting, reuse of illustrations, recitation, broadcasting, reproduction on microfilms or in any other physical way, and transmission or information storage and retrieval, electronic adaptation, computer software, or by similar or dissimilar methodology now known or hereafter developed.
The use of general descriptive names, registered names, trademarks, service marks, etc. in this publication does not imply, even in the absence of a specific statement, that such names are exempt from the relevant protective laws and regulations and therefore free for general use.
The publisher, the authors, and the editors are safe to assume that the advice and information in this book are believed to be true and accurate at the date of publication. Neither the publisher nor the authors or the editors give a warranty, expressed or implied, with respect to the material contained herein or for any errors or omissions that may have been made. The publisher remains neutral with regard to jurisdictional claims in published maps and institutional affiliations.

This Springer imprint is published by the registered company Springer Nature Singapore Pte Ltd.
The registered company address is: 152 Beach Road, #21-01/04 Gateway East, Singapore 189721, Singapore

Paper in this product is recyclable.

Contents

Editors and Contributors

About the Editors

Gavin Brett Melles Ph.D. (Education), M.Sc. (Sustainable Development), is a professor (adjunct) of sustainability and social innovation at Swinburne University's Design School. He was recently (2021–2022) a senior international fellow in sustainability at the Technical Design Department of Technical University Dresden. He is a research affiliate at the Centre for Design Innovations and Manufacturing Futures Research Institute at Swinburne University. His research and teaching interests include education, sustainability, and related issues.

Christian Wölfel Ph.D., is a trained industrial designer, researcher, and lecturer at the Chair of Industrial Design Engineering at Technische Universität Dresden, Germany. His research relates to the design and experience of complex systems in professional domains. It is grounded in cognitive and social sciences and aims for sustainable solutions. He teaches design fundamentals, design theory, and design research in interdisciplinary settings. He is a board member of the German Society for Design Theory and Research (DGTF) and works occasionally as a consultant for B2B companies.

Contributors

Ralph Boch Hans Sauer Foundation, Munich, Germany

Michael Burmester Hochschule der Medien Stuttgart, Stuttgart, Germany

Nina Costa Research Institute of Design, Media and Culture (ID+), University of Aveiro, Aveiro, Portugal

Harald Desing Empa, Technology and Society Laboratory, St.Gallen, Switzerland

Nadja Hempel Hans Sauer Foundation, Munich, Germany; Leuphana University (formerly), Lueneburg, Germany

Melanie Jaeger-Erben Hans Sauer Foundation, Munich, Germany; BTU Cottbus-Senftenberg, Cottbus, Germany

Linda Kästner Technische Universität Dresden, Dresden, Germany

Stefan Kaufman Monash University, Sydney, Australia

Anastassija Konash Fulbright Coral Sea Scholar Placement, Golisano Institute for Sustainability, Rochester Institute of Technology, Rochester, NY, USA

Jennifer Macklin Monash University, Melbourne, Australia

Gavin Brett Melles Faculty of Architecture and Industrial Design, Swinburne University, Melbourne, Australia;
Technische Universität Dresden, Dresden, Germany

Nicola Morelli Aalborg University, Copenhagen, Denmark

Nikolas Neumann Technische Universität Dresden, Dresden, Germany

Christian Nygaard Centre for Urban Transitions, Swinburne University of Technology, Melbourne, VIC, Australia

Christoph Scope Technische Universität Dresden, Dresden, Germany

Olamide Shittu Circular Economy Research Fellow, Centre for Urban Transitions, Swinburne University of Technology, Hawthorn, VIC, Australia

Luca Simeone Aalborg University, Copenhagen, Denmark

Ursula Tischner econcept–Agency for Sustainable Design, Cologne, Germany;
Wilhelm Büchner Mobile University of Technology, Darmstadt, Germany

Anne Velenturf School of Civil Engineering, University of Leeds, Leeds, England

Christian Wölfel Technische Universität Dresden, Dresden, Germany

Designing the Sustainable Circular Economy: From Products to Politics

1

Gavin Brett Melles and Christian Wölfel

1 Circular Economy in Context

Climate change demands global shifts away from fossil fuels and respect for development within safe ecological limits (Meadows et al. 2004; Rockström and Steffen 2009). As the most recent expression *Our Common Future* (World Commission on Environment and Development (WCED) 1987), the 17 Sustainable Development Goals (SDG) goals, targets and actions attempt to formulate processes by which growth might be sustainably achieved. In particular, business and industry are invited to focus on the goal SDG12 of sustainable consumption and production including through decoupling resource use and emissions from growth (Fisher 2020). The circular economy discourse has become in many jurisdictions, including the European Union (EU), the answer to achieving SDG12 and in many cases supplanting sustainable development as the goal itself (Harris et al. 2021). In so doing, the broader scope of systems change required by sustainable development has been unjustly forgotten or actively displaced (Blum et al. 2020).

Mainstream sustainable development is based on a mixture of market environmentalism (Fox et al. 2006), ecological modernisation (Pepper 1999), and populist discourses about human-nature connections (cf. Ives et al. 2017). Sustainability as a process is largely driven by technological innovation, green growth, voluntary regulation, and markets (Adams 2009). In contrast to this optimistic technical narrative, strong sustainable development is an agenda of socio-economic reform respecting ecological limits (Neumayer 2003), which envisages stronger regulation

G. B. Melles (✉)
Swinburne University, Melbourne, Australia
e-mail: gmelles@swin.edu.au; gavin.melles@msx.tu-dresden.de

G. B. Melles · C. Wölfel
Technische Universität Dresden, Dresden, Germany

© The Author(s), under exclusive license to Springer Nature Singapore Pte Ltd. 2024

G. B. Melles and C. Wölfel (eds.), *Design for a Sustainable Circular Economy*, Design Science and Innovation, https://doi.org/10.1007/978-981-99-7532-7_1

and policy, ecological economics, and an inclusive sustainability transition (Baker 2013).[1] The circular economy approach is at home in the mainstream sustainable development legacy, while scepticism about the capacity of technical and business innovation to promote growth finds expression in contrasting circular society narratives (e.g. Millar et al. 2019).

Given the dramatic nature of climate change emissions, a mainstream approach to sustainability is no longer a viable discourse for sustainability transitions. Hence, reconciling continued growth with reduced resource use and ecological thresholds appears to lead to contradictions in emissions (Hickel 2019). Far more ambitious approaches to change are required to achieve an ecologically sustainable transition for the planet (Soergel et al. 2021). Unless a multi-solving logic which identifies the interactions among social, environmental, and economic goals is adopted—the SDG project is likely to be fragmented and unsuccessful (Nilsson et al. 2016). This is where the idea of a sustainable circular economy and circular society can show what requirements are needed for real change.

2 Circular Economy as Engine of Change?

The circular economy model has been strongly promoted in Europe, China and other regions as the answer to green growth and transformation to a new socio-technical regime (Domenech and Bahn-Walkowiak 2019; Leipold 2021; Mathews and Tan 2016; Rudolph 2018). The EU sees the Circular Economy as a key mechanism in all sectors for the EU Green Deal and climate neutrality by 2050 (Rofifah 2020).

It is a familiar narrative of ecological modernisation (Christoff 1996). In its popular form, as adopted by governments and other actors, it offers an appealing rationale for business opportunity and industrial production and consumption decoupled from resource use and waste (Ellen MacArthur Foundation 2013). It is a model which builds on a history of discussion about minimising resource use, decoupling resource use from growth, and creating a more sustainable production and consumption system or performance economy[2] (Stahel 2020). However, latest independent reports suggest, the approach, has had a sustained global uptake of less than 9% (Morató et al. 2019; Platform for Accelerating the Circular Economy (PACE) 2020; de Wit 2019).

Criticisms of the concept as an umbrella term (Homrich et al. 2018) point to its origins in a variety of scientific, e.g. industrial ecology, and semi-scientific concepts and as a depoliticization of key socio-economic factors of sustainable

[1] Baker's Ladder of Sustainable Development spells out the multi-dimensional implications of the continuum of discourses or paradigms, including strong and weak sustainability.

[2] Stahel (2010) links this notion with a dematerialized economy with no waste streams—recycling being excluded—and focused on meeting human needs in an economically feasible way, including through carbon taxation and other resource redistribution tactics. In this way, a factor 10 reduction in consumption and GHG emissions will be achieved.

growth (Corvellec et al. 2021). Basic assumptions concerning the values, societal structures, cultures, underlying worldviews, and the paradigmatic potential of CE also remain largely unexplored (Korhonen et al. 2018). Critics have also identified theoretical and practical weaknesses in the mainstream proposals, including the rebound effect of increased consumption (e.g. Makov and Vivanco 2018), and a disconnect from sustainable development (Garcia-Muiña et al. 2019).

Although protagonists see the spread and popularisation of the model as the goal—circularity for circularity's sake (Harris et al. 2021), any assessment of its value must be measured by its contribution to sustainable development (Blum et al. 2020). Indeed, a circular economy (and society) can only be considered sustainable, when it leads to respecting planetary boundaries (Desing et al. 2020). With reference to the SDGs, the aim of sustainable consumption and production is of relevance to the approach. Several other goals, e.g. SDG 7 Affordable and Clean Energy, identify elements which also can play a role, especially in the socio-technical realm of the economy (Ghosh 2020).[3] However, if the circular economy is to contribute to these goals several theoretical, semantic and practical weaknesses in the current formulation need repairing (Millar et al. 2019).

Given its limited global uptake and the relatively conservative nature of the changes, e.g. increased recycling rates, product longevity, etc., the mainstream circular economy could be seen as an 'adjustment' rather than a disruptive change. Circular economy assumes a mainstream neo-liberal economy and does not address how markets, society, government, and the environment interact (Stegeman 2015). Thus it 'lacks an economic theory that can pragmatically guide the transition from the prevailing neoclassical model towards one that would drive the transition towards a sustainable circular economy and be palatable for governments' (Velenturf and Purnell 2021, p. 1453).

3 Multi-level Sustainability Transition(s)

One striking weakness of discussions of circular economy is that its implementation is seen as the goal itself rather than a means to an end—sustainable development. Multi-level perspective on sustainability transitions theory is a 'middle-range framework' (Geels 2011) for a new configuration of production and consumption processes in contexts of broader socio-economic and political change. Transitions theory has strong support from the EU as a guide to the promotion of sustainability through government policy and other mechanisms (European Environmental Agency (EEA) 2019). Multi-level perspective on sustainability transitions offers a vision of the ecological, institutional, and technical drivers which

[3] A fact which is discussed by Sutherland & Koulumpi in IISD Blog.
https://sdg.iisd.org/commentary/guest-articles/more-than-just-sdg-12-how-circular-economy-can-bring-holistic-wellbeing/

can promote change (Loorbach et al. 2017).[4] As a middle-range theory (cf. Merton 1968) it is centrally concerned with socio-technical regime change—a key dimension of overall sustainability transition—and hence it has been linked to circular economy (Jackson et al. 2014; Oliveira et al. 2021).

As Geels (2011) has shown, socio-technical regime changes depend on bottom-up, e.g. niche innovations in protected spaces, and top-down initiatives in policy, science and socio-cultural changes in the context of changing 'landscape' conditions and pressures. A new socio-technical regime emerges because landscape changes and disruptions destabilise the status quo and allow new policy and economic configurations to emerge, which themselves bring (existing) niche innovations into the foreground, e.g. renewables and new zero energy sources, such as green hydrogen. At the micro-level niche, innovations visualise new business—society—environment constellations, e.g. the sharing economy, and possibilities, hitherto marginal, align and scale with these changes to become part of a new regime (Augenstein et al. 2020; Hargreaves et al. 2013; Smith 2004).

While transitions theory tends to represent socio-technical change as a change from a current to *the* new regime, as Paredis (2011) notes there are multiple possible sustainability transitions, underpinned by different scenarios of what this can look like (see Olson 1995). Transition visions depend on the details of the new economy in contrast to business-as-usual neo-liberal growth (Longhurst et al. 2016). For example, a transition landscape of social solidarity, appropriate technology and ecological limits will look very different as a socio-technical regime changes to one where markets for green growth under technical optimism take centre stage. So, it is important that socio-technical regime change is not viewed as singular (Genus 2014). This becomes important when assessing the principles and goals of competing circular narratives, policies, and politics (Leipold et al. 2021).

Typical government-industry reactions are new policies, standards, and norms for the socio-technical realm. We show in what follows that the mainstream circular economy narrative is only one of several narratives including a role for socio-technical regime change. We also note that it is a modest incremental change to the status quo, has been hitherto relatively unsuccessful in implementation, and requires a significantly expanded agenda to achieve inclusive development for the future (Fig. 1).

4 Circular Economy Discourses

The articulation of a sustainable circular economy is to specify the elements of this theory of transition in recognition of the fact that there are competing circular discourses in many countries (e.g. Melles 2021). Thus, Friant and Vermeulen

[4] While Loorbach et al. (Loorbach et al. 2017) suggest that socio-technical, socio-ecological and socio-institutional perspectives differentiate the field, these appear to us to be all elements of the multi-level perspective, which can be fore- or backgrounded.

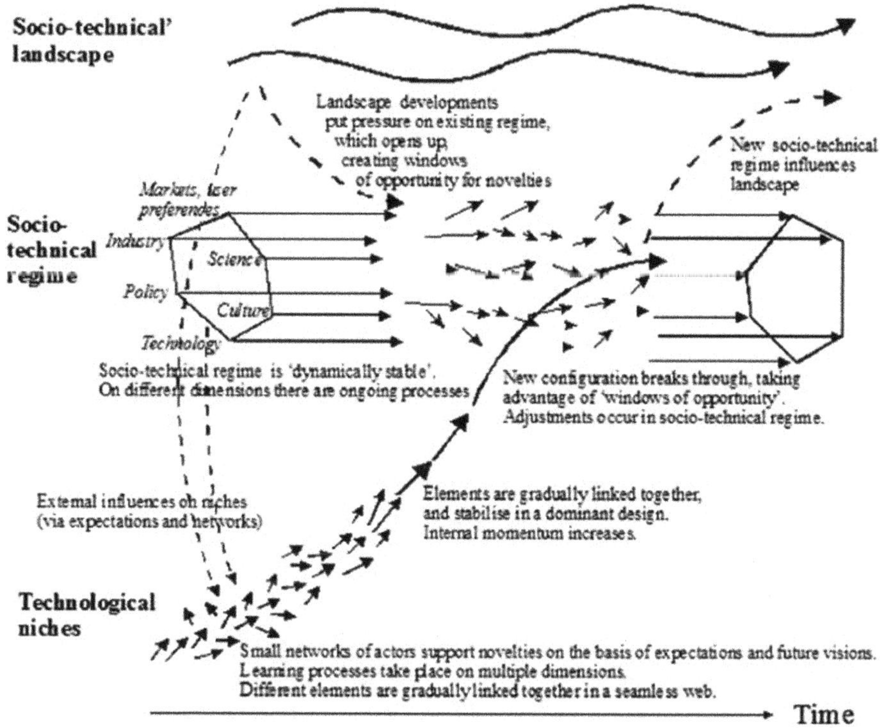

Fig. 1 Sustainability transitions (Geels 2002)

(2020a) develop a typology of four broad circularity visions relevant to sustainable development. In their framework, circular economy is described as a relatively conservative techno-centric circular discourse (Calisto Friant et al. 2020b), and compared to other more reformist and transformational circular society proposals, e.g. Doughnut Economics (Raworth 2019). The nature of the niche innovations and higher-level regime changes in business, industry, policy etc., are consistent with this non-disruptive socio-technical 'imaginary' (Kovacic et al. 2019).

Most relevant to this discussion is the contrast of the mainstream 'technicist' with the reformist circular economy, including with reference to design. Both narratives see a role for capitalism, markets and growth albeit under degrees of revision consistent with the widely circulated contrast of weak and strong sustainability around growth limits and conservation of natural capital (Melles et al. 2015; Williams and Millington 2004).[5] In comparison to the mainstream circular

[5] They cite Doughnut Economics (Raworth 2017) as an example of a reformist agenda although a close reading of Raworth suggests this model has elements of an even more sceptical approach to capitalism. Thus, their claim that reformist models fail to address questions of economic growth, entropy and decoupling (Calisto Friant et al. 2020a, p. 11) is incorrect.

economy model, the circular society model adds considerations of participation and equitable resource distribution (Jaeger-Erben et al. 2021).

The combination of a circular economy and circular society model is consistent with what elsewhere is described as a sustainable circular economy (Velenturf et al. 2019). The society model places emphasis on participatory and co-design approaches to the social innovation required to promote system change (see Britton 2017). It is a narrative that continues to see a role for growth and markets but not without social and political reform of an economy operating within ecological boundaries. Hence, circular economy and society suggest that material and political changes are needed and hence embrace both expert and diffuse design approaches for social innovation (Manzini 2015).

5 Design for Circular Economy and Society

The role of design has been increasingly highlighted as 'catalyst' (Moreno et al. 2016) for the circular economy. The importance of design for sustainability is a discussion with a long history that predates recent circular economy and design discussions. An early critique of the unsustainability of expert industrial design was the work of Viktor Papanek—Design for the Real World (Papanek 1971), which gave rise to the development of social design (Melles et al. 2011). The expansion of industrial design and other design fields into co-design and co-creation services and other spaces, e.g. policy, outside of expert design per se (Sanders and Stappers 2008), has meant that the question of design's contribution to sustainable development has become more complex (Ceschin and Gaziulusoy 2016).

Sustainable development requires design to integrate systems thinking in multiple ways. In order to promote more radical changes on industrial and societal scope, e. g. by implementing sufficiency strategies and changing consumption patterns accordingly, product-service systems design has been identified as a compelling approach (e. g. 2021; Velenturf and Purnell 2021). However, complex changes in product and service offerings yield the risk to fail if these are not aligned with policy on one hand or the consumer perspective on the other hand.

Participatory approaches have been suggested to address such discrepancies and success barriers (Lofthouse and Prendeville 2018; Velenturf and Purnell 2021). Co-designing and other forms of participatory development serve not only as an approach to meet expectations and hence gain acceptability among consumers and citizens. If implemented accordingly, co-design can help to promote inclusive development, hence fostering social justice in sustainability transition (cf. Melles 2019). This is consistent with the Sustainable Development Goals and notions of a sustainable circular economy and is hence advocated by corresponding strategies and roadmaps (Melles et al. 2022). Still, there are methodical challenges seen on how to integrate the multitude of stakeholders in co-creation (Buhl et al. 2019; Kagan et al. 2019). Despite extensive efforts on integrating user and citizen perspectives in designing for sustainable circular economy, "most of […] circular economy seems to focus on the production side", while less attention is being paid to consumers (Camacho-Otero et al. 2018). While contemporary human-centred design offers the approaches and means to empathize with users, to take their perspective and to pay regard to their needs, this is often not applied here. Among the major factors that promote or prevent the acceptance and hence the success of circular design solutions among customers are psychological factors such as personal characteristics, expected risks and uncertainty or attitudes and habits (ibid.). Participatory approaches can address some of these factors, and specific design principles are available that address some of these factors (e. g. Wallner et al. 2022). However, in order to systematically address these psychological success factors, respective human-centred expert design approaches need to be included.

The traditional product-related design focus still largely defines the practice of industrial design. A helpful distinction between this and the broader non-design focused applications of design into policy (e.g. Howlett 2020) and other spaces is that of Manzini (2015), who in the context of social innovation distinguishes expert and diffuse design—hybrids of expert design in diffuse design contexts are also possible. Thus, Andrews (2015) and others are correct in saying design has an influential role in circular economy and sustainable development but without a clear articulation of which circularity—technicist or society-oriented—and whether in expert and diffuse domains, current images of designs contribution are inadequate.

6 Principles for a Sustainable Circular Design

"*To design is to devise courses of action aimed at changing existing situations into preferred ones.*" (Herbert Simon 1996).

Velenturf and Parnell (2021) define the challenge of reconnecting circular economy with sustainable development,

> Circular economy must be fully integrated with sustainable development. This necessitates a profound reconsideration of circular economy, broadening its scope from closed-loop recycling and short-term economic gains, towards a transformed economy that organises access to resources to maintain or enhance social welbeing and environmental quality. Superficial changes, i.e. to accommodate recycling, to prevailing economic models will not suffice (p. 1453).

They identify ten principles which describe how to mobilise society, business and industry, and government to develop sustainable circular economy. This expanded model to include multiple government, industry and other stakeholders is not included in standard narratives.

1. Beneficial reciprocal flows of resources between nature and society
2. Reduce and decouple resource use
3. Design for circularity
4. Circular business models to integrate multi-dimensional value

5. Transform consumption
6. Citizen participation in sustainable transitions
7. Coordinated participatory and multi-level change
8. Mobilise diversity to develop a plurality of circular economy solutions
9. Political economy for multi-dimensional prosperity
10. Whole system assessment

This proposal, which is referenced by several of the authors of this book, and which is taken up and developed in chapter 7 below by Melles and Velenturf, constitutes the network of principles that define for this book sustainable circular design. In so doing, we propose a unique definition of design and see our proposal as the engine of change for sustainability transitions. We see similarities between our proposal and other positions such as those of Doughnut Economics, Circular Society and related frameworks. We differ in proposing that design in both a material and political sense is required for the wholesale systems change required.

7 Conclusion

In this introductory chapter, we have examined the argument for circular economy as the engine of sustainability transition and observed that such a position under-represents the circular discourses and possibilities for change. Secondly, we agree with scholars who see the need for a re-imagining of the contribution of design to this project (De los Rios and Charnley 2017) but our response is distinct in seeing an expert and diffuse design at work in creating a sustainable circular economy.

The mainstream circular economy discourse assumes that green growth can be achieved through the 'debatable' process of decoupling production and consumption from growth (Fletcher and Rammelt 2017). We also noted that the mainstream story fails to critique the premise of continuous growth as consistent with sustainable development (e.g. Corvellec et al. 2021). Without clear articulation of principles for a sustainable circular economy, key aspects of change are absent.

Circular Design we have suggested must not be limited to a set of industrial design techniques, e.g. design for remanufacture or modularity, but must include policy design. We concur with Kovacs et al. who concludes that 'Circular economy policies would be a success even though the economy cannot be circular, if they could inspire and stimulate creativity and entrepreneurship in civil society to develop and prepare steppingstones and building blocks towards a type of civilisation that destroys less of the biosphere' (Kovacic et al. 2019, p. 169).

The editors, our co-authors and a growing number of scholars share a concern with the narrow depiction of circular economy as an adequate engine for sustainability transition. We also share together a belief that circular design as both a diffuse and expert discipline has a key role to play in influencing the establishment of a sustainable circular economy. To be successful in this urgent enterprise we need to heed the findings and insights described in these chapters and beyond, and engage with the challenge through a broad circular design of products, industry, society and politics.

8 Author Contributions

With a view to developing a clearer understanding of the implications of an appropriately broad account of circular design (integrating both expert and diffuse design perspectives) as the guiding framework for a sustainable circular economy and transition to a circular society, we (the editors) identified leading scholars who shared a scepticism towards the technicist narrative of circular economy and had recently research and published on these issues. This process of identifying co-authors took place while one of the editors (Melles) was undertaking a senior international fellowship in Sustainability at the Chair of Industrial Design Engineering at TU Dresden, Germany through 2021—2022. The fellowship has been funded by the German Federal Ministry of Education and Research and the Free State of Saxony through the Excellence Strategy of the German federal and state governments. The editors together are grateful for the support in helping make this collaboration possible.

All the authors share some commitment albeit with nuances towards a sustainable circular economy and society transition and have beliefs about the signal role of design—both expert and diffuse—in achieving such a transition. As illustrated in their chapters later in this book, it also becomes clear that this particular network of fellow travellers subscribes to one or more of the ten principles above (Velenturf and Purnell 2021). Chapter 7 by Melles and Velenturf takes up the original proposition and expands on its implications for industry and society.

In addition to this broad shared basis, all the included authors then bring to the debate about a circular sustainability transition, specific new perspectives, and arguments to this more rigorous and critical challenge to the mainstream narrative of circular economy. Scholars and practitioners seeking guidance on the challenges to the mainstream belief in technical solutions to green growth will find valuable material here. In what follows we outline briefly what these different contributions are and provide a brief profile of the authors from Europe and Australia who have brought a wealth of background, experience and focus to the topic of this book. Below we provide a short summary of the key facets of the chapters to come.

Chapter 1: Designing a Sustainable Circular Economy, *Gavin B. Melles and Christian Wölfel*

In this introduction, we introduce the frameworks, principles and theories which inform our approach to a sustainable circular economy. We identify the ten principles of a sustainable circular economy and suggest how they might link to a hitherto new model of circular design. This model of circular design includes both expert industrial design considerations and diffuse social and policy design considerations.

Chapter 2: Beyond Eco-Design to Circular Design, *Ursula Tischner*

Societies in industrialised countries and economies in transition have developed lifestyles based on resource consumption levels that cannot be maintained by most of the world's population without ruining our natural environment. Therefore, consumers and producers need to re-imagine new ways of sustainably producing and consuming. Increasing resource efficiency and introducing a more circular economy are important elements of this shift. This chapter explores how this change might be supported by design combining findings from environmental and behavioural science, happiness and design research. Starting with a discussion of the limits to resource consumption and a critical view on current lifestyles, the quest for happiness and wellbeing as well as the circular economy and the roles of designers some new approaches are presented that shift the focus from circular design towards the design of sustainable circular systems. Finally, a model is presented of how designers can help to make more sustainable production-consumption systems desirable based on findings from behavioural psychology.

Chapter 3: Connecting Global Sustainability with Circular product Design, *Harald Desing*

The circular economy has become popular among private and public actors as a solution to these multiple environmental crises. However, this promise may only lead to environmental sustainability, if the biophysical limits of our planet are respected. This chapter introduces the'resource pressure ' method, which aims at guiding design decisions by connecting global sustainability criteria with resource effectiveness on a product level. First, Earth system boundaries are translated into resource budgets, i.e. the maximum annual production of materials and energy, which are environmentally sustainable. These resource budgets are then used as a reference for the resource consumption induced by the product or service, i.e. the'pressure ' induced on limited resources. Circular strategies, such as optimising lifetime, remanufacturing, recycling and cascading, can be evaluated case-by-case on their effect on resource consumption and the reduction in resource pressure. Targets for reduction of resource consumption can be derived and their achievement measured.

Chapter 4: Mapping Circular economy and the Role of Design in Portugal, *Nicola Morelli, Nina Costa and Luca Simeone*

Although the circular economy has become a "hot" topic for discussion in the last decade, it represents the maturation of different streams of research on environmental strategies and sustainability in the last 30 years. The national context analyzed in this paper is Portugal. The authors map the current state of circularity in Portugal and identify the scope of application. In their conclusion, they note the need for deeper engagement. The call to action in this chapter should be read in relation to the other chapters outlining the nature of a sustainable CE and society, as well as compared to the circular state of play in other European countries.

Chapter 5: Circular Packaging in the Cosmetics Industry—A Systematic Review on Challenges and the Current State of Sustainable Strategies and Solutions, *Christoph Scope, Nikolas Neumann, Christian Wölfel, Linda Kästner*

The largest share of plastics is used for packaging of goods (almost 40% in the EU), resulting in a correlating impact on resource use, carbon emissions and landfill. The application of sustainable strategies to packaging should help to reduce negative impacts and promote the development of sustainable consumption. The authors of this chapter set out the need for a fundamental change in cosmetics packaging with its exacerbated requirements, which is already mirrored by relevant regulations that will take effect soon. Established cosmetics packaging design strategies do not meet these requirements. A shift to sustainable circular packaging is needed in the sector. A systematic literature review is conducted that analyses the body of research on circular packaging design in the cosmetics industry. The authors identify current challenges for sustainable packaging that are specifically exacting in the cosmetics sector e. g. due to functional and hygienic requirements. In addition to the functions and current standards of packaging, packaging alternatives are presented, which are categorised according to the R-strategies Refuse, Reduce, Reuse and Recycle. Subsequently, the aspects of customer acceptance and decision-making are discussed as relevant factors for the success of circular packaging solutions in the cosmetics industry. It is found that there is a gap between known sufficiency strategies of sustainable circular economy and its implementation in cosmetics packaging, as it is in packaging in general.

Chapter 6: Systems Perspectives on Circular Economy for the Design in Manufacturing, *Anastasia Konash*

Circular economy is often seen as a way to deal with waste and overconsumption and achieve sustainable living. When discussing the implementation of circular economy, manufacturers tend to be perceived as gatekeepers to more sustainable and more circular products as well as the main polluters. They also are identified as the main beneficiaries of new circular business models. However, in reality, circular economy and manufacturing are only part of what is needed to achieve the sustainability transition we need, and both managers and policy makers tend to overlook the broader system in which they function. Addressing this overlook, in this chapter the authors review the opportunities and limitations of circular economy as they are perceived by manufacturers. Coming on the back of decades of efficiency measures and sustainable manufacturing campaigns, they look at the drivers and challenges which manufacturers face while implementing sustainability measures. The authors discuss the necessity of modifying business models and briefly review the power of manufacturers in changing the design and longevity of the products. They draw the conclusion that wider societal changes in consumption are required to achieve the sustainable circular economy goals.

Chapter 7: Circular Design for a Transition to a Sustainable Circular Society: Defining a New Profession, *Gavin B. Melles and Anne Velenturf*

Mainstream circular economy emphasizes the closing of material loops as the way to ensure green growth, and there is a key role for design to achieve such change. According to reports, however, global appetite for circular economy remains limited and critics of mainstream circularity have pointed to several contradictions between the rhetoric and reality of circular change. Circular economy is one of several discourses about socio-technical regime range required for a sustainability transition. The circular society narrative for example sees reform of current economic thinking towards more inclusive models and agnosticism about the imperative of economic growth. In both models, product design and design of policy are important for promoting sustainable change. However, current formulations of circular economy misrepresent the plurality of circular discourses and hence the role of expert and diffuse design. In this chapter, the authors compare the circular economy and circular society arguments for sustainability transition and employ the ten principles of a sustainable circular economy to articulate a broader circular design agenda than typically proposed.

Chapter 8: Practice Perspective Implications for Sustainable Circular Economy transitions, *Olamide Shittu and Christian Nygaard*

The discourse on sustainable CE transitions stems from the need to provide sustainable solutions to global environmental and resource use challenges. However, the understanding of sustainable CE practices remains emergent as we cannot easily identify the materials, meanings and competences that constitute a coherent sustainable CE practice. Practice theory calls for a move beyond the extant focus on structures or individual actions in sustainability transitions by centralising the bundle of activities that order and give meaning to social living. Focusing on household plastic consumption, the doctoral research discussed in this chapter explored how to enable the transition of cities to sustainable CEs by transforming day-to-day practices. Hence, the authors provide an alternative analytical framework of sustainable CE practices existing on three different variations and stages of materialisation. These are input assemblage, input combinations, and outputs and outcomes. Lastly, we contend that transitioning to a sustainable CE entails going beyond business-as-usual to re-imagine new, radical, and disruptive practice configurations in socio-technical systems.

Chapter 9: Addressing Psychological Needs in Designing for a Sustainable Circular Economy, *Christian Wölfel and Michael Burmeister*

Approaches and methods on designing for a sustainable circular economy are developing solutions for sustainable development. For such solutions to succeed, market and social acceptance of circular products and services must improve. Currently, there is a mixed reception among consumers and hence a stagnating market share of circular solutions. In this chapter, the authors discuss these limitations and relate them to psychological aspects of consumer and user experience and behaviour. Research on user experience and experience design has delivered theories, approaches and methods on how to design for people as they experience and behave in the world. One core finding is that psychological needs play an important role in human-product interaction. The fulfilment of such needs results in positive experiences and can raise acceptance of products and services. The authors illustrate how psychological needs can be addressed in a sustainable circular design. They present a fictitious case to illustrate how specific non-instrumental qualities of offerings can address psychological needs and hence fundamentally influence overall judgements and behaviour in interaction with sustainable circular solutions. Finally, the authors discuss how such needs-based experience design can be implemented in design processes.

Chapter 10: Blending Design and Behavioural Science in Three Linked Public Policy Experiments towards a Circular Economy, *Stefan Kaufman, Jennifer Macklin and Sebastian Jarvol*

People's behaviour, choices and desires are increasingly foregrounded in the challenge of decoupling the material and energy footprint of everyday life from wellbeing and happiness. Two leading perspectives on people and behaviour in public policy and administration are human-centred design and Behavioural Insights. A common dichotomy contrasts Behavioural Insights as at best incremental and human-centred design as participatory emancipation. However, a more nuanced, balanced, and integrated body of theory and practice is possible blending both design and behavioural science experimentation which the authors label Behavioural Public Policy and Administration. This chapter reflects on three linked corresponding interventions on household recycling contamination, consumer eco-labelling, and business innovation for circularity. The authors argue for the value of a focus on behaviour as opening the potential for integration and complementarity between jurisdictions, stakeholders and disciplinary perspectives, while also exploring the implications of findings from our projects for the transition to a CE.People's behaviour, choices and desires are increasingly foregrounded in the challenge of decoupling the material and energy footprint of everyday life from wellbeing and happiness. Two leading perspectives on people and behaviour in public policy and administration are human-centred design and Behavioural Insights. A common dichotomy contrasts Behavioural Insights as at best incremental and human-centred design as participatory emancipation. However, a more nuanced, balanced, and integrated body of theory and practice is possible blending both design and behavioural science experimentation which the authors label Behavioural Public Policy and Administration. This chapter reflects on three linked corresponding interventions on household recycling contamination, consumer eco-labelling, and business innovation for circularity. The authors argue for the value of a focus on behaviour as opening the potential for integration and complementarity between jurisdictions, stakeholders and disciplinary perspectives, while also exploring the implications of findings from our projects for the transition to a CE.

Chapter 11: Co-designing a Circular Society, *Nadja Hempel, Ralph Boch and Melanie Jäger-Erben*

The majority of political, scientific and economic measures for a transition towards Circular Economy focuses on technological and business model innovations. They largely exclude societal transformation efforts and socially innovative change attempts. Issues of improving quality of life, promoting sufficiency, changing social inequalities and unequal power relations in production and consumption systems are mainly addressed in contributions that provide a critical perspective on circular economy. However, the Circular Economy concept has the potential to become a comprehensive social-ecological transformation program if these issues are consistently included. The term Circular Society has been introduced by a diverse range of actors from science, economy and civil society to provide a complementary or alternative framing to circular strategies. Yet the field of research and practice developing under this term is still in its infancy, different strands of discussion have not been explored and synthesised yet. This chapter presents the process and its results so far and reflects upon methodological learnings and insights on key circular society principles, actors, and possible conflict issues. Furthermore, by working and experimenting with a combination of transformative and design-oriented research approaches, methodological and epistemological contributions to scientific practices for sustainability are made.

References

Adams WM (2009) Green development: environment and sustainability in a developing world, 3rd edn. Routledge Taylor & Francis Group, Abingdon

Andrews D (2015) The circular economy, design thinking and education for sustainability. Local Econ 30(3):305–315. https://doi.org/10.1177/0269094215578226

Asia-Pacific Economic Cooperation (2020) Circular economy roadmap—Malaysia's practice (2020/SOM1/PPSTI/029; Issue February).http://mddb.apec.org/Documents/2020/PPSTI/PPSTI1/20_ppsti1_029.pdf. Accessed 5 Apr 2023

Augenstein K, Bachmann B, Egermann M, Hermelingmeier V, Hilger A, Jaeger-Erben M, Kessler A, Lam DPM, Palzkill A, Suski P, Von Wirth T (2020) From niche to mainstream: the dilemmas of scaling up sustainable alternatives. Gaia 29(3):143–147. https://doi.org/10.14512/GAIA.29.3.3

Baker S (2013) Sustainable development. Routledge, London

Blum NU, Haupt M, Bening CR (2020) Why "Circular" doesn't always mean "Sustainable." Resour Conserv Recycl 162(June):105042. https://doi.org/10.1016/j.resconrec.2020.105042

Bocken NMP, de Pauw I, Bakker C, van der Grinten B (2016) Product design and business model strategies for a circular economy. J Ind Prod Eng 33(5):308–320. https://doi.org/10.1080/21681015.2016.1172124

Boxall NJ, King S, Kaksonen A, Bruckard W, Roberts D (2019) Waste innovation for a circular economy: a summary report for the CSIRO Cutting Edge Science and Engineering Symposium 2019 (Issue August).https://publications.csiro.au/rpr/pub?pid=csiro:EP195506

Britton G (2017) Co-design and social innovation: connections, tensions and opportunities. (P. (Firm) issuing body (ed.), 1st edn. Routledge, London

Bruntland G (1987) Report of the World Commission on Environment and Development: "our common future". Oxford University Press, Oxford

Buhl A, Schmidt-Keilich M, Muster V, Blazejewski S, Schrader U, Harrach C, Schäfer M, Süßbauer E (2019) Design thinking for sustainability: why and how design thinking can foster sustainability-oriented innovation development. J Clean Pro 231:1248–1257. https://doi.org/10.1016/j.jclepro.2019.05.259

Calisto Friant M, Vermeulen WJV, Salomone R (2020a) A typology of circular economy discourses: Navigating the diverse visions of a contested paradigm. Resour Conserv Recycl, 161:104917. https://doi.org/10.1016/j.resconrec.2020.104917

Calisto Friant M, Vermeulen WJV, Salomone R (2020b) A typology of circular economy discourses: navigating the diverse visions of a contested paradigm. Resour Conserv Recycl 161(April):104917. https://doi.org/10.1016/j.resconrec.2020.104917

Camacho-Otero J, Boks C, Pettersen IN (2018) Consumption in the circular economy: a literature review. Sustainability (Switzerland) 10(8). https://doi.org/10.3390/su10082758

Ceschin F, Gaziulusoy I (2016) Evolution of design for sustainability: from product design to design for system innovations and transitions. Des Stud 47:118–163. https://doi.org/10.1016/j.destud.2016.09.002

Christoff P (1996) Ecological modernisation, ecological modernities. Environ Polit 5(3):476–500. https://doi.org/10.1080/09644019608414283

Circular Economy Initiative Deutschland (2021) Circular-Economy-Roadmap-fuer-Deutschland

Commonwealth Government. (2018) 2018 National waste policy: less waste, more resouces. https://doi.org/10.4337/9781784715885

Corvellec H, Stowell, AF, Johansson N (2021) Critiques of the circular economy. J Ind Ecol, 1–12.https://doi.org/10.1111/jiec.13187

De los Rios IC, Charnley FJS (2017) Skills and capabilities for a sustainable and circular economy: The changing role of design. J Clean Prod 160:109–122.https://doi.org/10.1016/j.jclepro.2016.10.130

den Hollander MC, Bakker CA, Hultink EJ (2017) Product design in a circular economy: development of a typology of key concepts and terms. J Ind Ecol 21(3):517–525. https://doi.org/10.1111/jiec.12610

de Wit M, Verstraeten-Jochemsen J, Hoogzaad J, Kubbinga B (2019) The Circularity gap report 2019: closing the circularity gap in a 9% World. Circle Econ Amsterdam. Available at https://docs.wixstatic.com/ugd/ad6e59_ba1e4d16c64f44fa94fbd8708eae8e34.pdf. Accessed 01.11.23

Desing H, Braun G, Hischier R (2021) Resource pressure—a circular design method. Resour Conserv Recycl 164:105179. https://doi.org/10.1016/j.resconrec.2020.105179

Desing H, Brunner D, Takacs F, Nahrath S, Frankenberger K, Hischier R (2020) A circular economy within the planetary boundaries: Towards a resource-based, systemic approach. Resour, Conserv Recycl 155:104673. https://doi.org/10.1016/j.resconrec.2019.104673Domenech

Domenech T, Bahn-Walkowiak B (2019) Transition towards a resource efficient circular economy in Europe: policy lessons from the EU and the Member States. Ecol Econ 155(March 2017):7–19. https://doi.org/10.1016/j.ecolecon.2017.11.001

Ellen MacArthur Foundation (2013) Towards the ciruclar economy: economic and business rationale for an accelerated transition (No. 1; Issue 1).https://www.ellenmacarthurfoundation.org/assets/downloads/publications/Ellen-MacArthur-Foundation-Towards-the-Circular-Economy-vol.1.pdf

European Environmental Agency (EEA) (2019) Sustainability transitions: policy and practice.https://doi.org/10.2800/641030

Fisher B (2020) Time to revise the SDGs. Nature 583(July):331–332

Fletcher R, Rammelt C (2017) Decoupling: a key fantasy of the post-2015 sustainable development agenda. Globalizations 14(3):450–467. https://doi.org/10.1080/14747731.2016.1263077

Fox G, Anderson TL, Leal DR (2006) Free market environmentalism. Am J Agric Econ 74:500. https://doi.org/10.2307/1242506

Fudge C, Rowe J (2001) Ecological Modernisation as a framework for sustainable development: a case study in Sweden. 33(9):1527–1546.https://doi.org/10.1068/a33153

Garcia-Muiña FE, González-Sánchez R, Ferrari AM, Volpi L, Pini M, Siligardi C, Settembre-Blundo D (2019) Identifying the equilibrium point between sustainability goals and circular

economy practices in an Industry 4.0 manufacturing context using eco-design. Soc Sci 8(8). https://doi.org/10.3390/socsci8080241
Geels FW (2011) The multi-level perspective on sustainability transitions: responses to seven criticisms. Environ Innov Soc Trans 1(1):24–40. https://doi.org/10.1016/j.eist.2011.02.002
Geels FW (2002) Technological transitions as evolutionary reconfiguration processes: A multi-level perspective and a case-study. Res Policy 31:1257–1274. https://doi.org/10.1016/S0048-7333(02)00062-8
Genus A (2014) Governing sustainability: a discourse-institutional approach. Sustainability (Switzerland) 6(1):283–305. https://doi.org/10.3390/su6010283
Ghosh, S. K. (2020). Introduction to circular economy and summary analysis of chapters. In: Circular economy: global perspective. Springer, Singapore, pp 1–23. https://doi.org/10.1007/978-981-15-1052-6_1
Gibbs DC, Longhurst J, Braithwaite C (1998) "Struggling with sustainability": weak and strong interpretations of sustainable development within local authority policy. Environ Plan A 30(8):1351–1365. https://doi.org/10.1068/a301351
Halstenberg FA, Stark R (2019) Introducing product service system architectures for realizing circular economy. Procedia Manufacturing 33:663–670. https://doi.org/10.1016/j.promfg.2019.04.083
Hargreaves T, Hielscher S, Seyfang G, Smith A (2013) Grassroots innovations in community energy: the role of intermediaries in niche development. Glob Environ Chang 23(5):868–880. https://doi.org/10.1016/j.gloenvcha.2013.02.008
Harris S, Martin M, Diener D (2021) Circularity for circularity's sake? Scoping review of assessment methods for environmental performance in the circular economy. Sustain Prod Consum 26:172–186. https://doi.org/10.1016/j.spc.2020.09.018
Hickel J (2019) The contradiction of the sustainable development goals: Growth versus ecology on a finite planet. Sustain Dev 27:873–884. https://doi.org/10.1002/sd.1947
Homrich AS, Galvão G, Abadia LG, Carvalho MM (2018) The circular economy umbrella: trends and gaps on integrating pathways. J Clean Prod 175:525–543. https://doi.org/10.1016/j.jclepro.2017.11.064
Howlett M (2020) Challenges in applying design thinking to public policy: dealing with the varieties of policy formulation and their vicissitudes. Policy Polit 48(1):49–65. https://doi.org/10.1332/030557319X15613699681219
Ives CD, Giusti M, Fischer J, et al (2017) Human–nature connection: a multidisciplinary review. Curr Opin Environ Sustain 26–27:106–113. https://doi.org/10.1016/j.cosust.2017.05.005
Jackson M, Lederwasch A, Giurco D (2014) Transitions in theory and practice: managing metals in the circular economy. Resource 3:516–543. https://doi.org/10.3390/resources3030516
Jaeger-Erben M, Jensen C, Hofmann F, Zwiers J (2021) There is no sustainable circular economy without a circular society. Resour Conserv Recycl 168(January):105476. https://doi.org/10.1016/j.resconrec.2021.105476
Kagan C, Burton M, Duckett P, et al (2019) Critical community psychology: Critical action and social change. Second Edition
Kłosińska K (2019) Mapa Drogowa: Transformacji w kierunku gospodarki o obiegu zamkniętym. https://www.gov.pl/attachment/72d8cd08-f296-43f5-af28-21ab2fada40e
Korhonen J, Nuur C, Feldmann A, Birkie SE (2018) Circular economy as an essentially contested concept. J Clean Prod 175:544–552. https://doi.org/10.1016/j.jclepro.2017.12.111
Kovacic Z, Strand R, Völker T (2019) The circular economy in Europe: critical perspectives on policies and imaginaries. In: Routledge explorations in sustainabilty and governance (Issue November). Routledge Taylor & Francis Group. https://doi.org/10.4324/9780429061028
Lah O (2016) Circular economy policies and strategies of Germany. In: Anbumozhi V, Kim J (eds) Towards a circular economy: corporate management and policy pathways (Issue September). ERIA, pp 59–74
Leipold S (2021) Transforming ecological modernization 'from within' or perpetuating it? The circular economy as EU environmental policy narrative. Env Polit 30:1045–1067. https://doi.org/10.1080/09644016.2020.1868863

Leipold S, Weldner K, Hohl M (2021) Do we need a 'circular society'? Competing narratives of the circular economy in the French food sector. Ecol Econ 187:107086. https://doi.org/10.1016/j.ecolecon.2021.107086
Lofthouse V, Prendeville S (2018) Human-centred design of products and services for the circular economy–a review. Des J 21(4):451–476. https://doi.org/10.1080/14606925.2018.1468169
Longhurst N, Avelino F, Wittmayer J, Weaver P, Dumitru A, Hielscher S, Cipolla C, Afonso R, Kunze I, Elle M (2016) Experimenting with alternative economies: four emergent counter-narratives of urban economic development. Curr Opin Environ Sustain 22:69–74. https://doi.org/10.1016/j.cosust.2017.04.006
Loorbach D, Frantzeskaki N, Avelino F (2017) Sustainability transitions research: transforming science and practice for societal change. Annu Rev Environ Resour 42:599–626. https://doi.org/10.1146/annurev-environ-102014-021340
Makov T, Vivanco DF (2018) Does the circular economy grow the pie? The case of rebound effects from smartphone reuse. Front Energy Res 6(MAY):1–11. https://doi.org/10.3389/fenrg.2018.00039
Manzini E (2015) Design when everybody designs: an introduction to design for social innovation. MIT Press
Mathews JA, Tan H (2016) Circular economy: lessons from China. Nature 531(7595):440–442. https://doi.org/10.1038/531440a
McDowall W, Geng Y, Huang B, Barteková E, Bleischwitz R, Türkeli S, Kemp R, Doménech T (2017) Circular economy policies in China and Europe. J Ind Ecol 21(3):651–661. https://doi.org/10.1111/jiec.12597
Meadows D, Randers J, Meadows D (2004) Limits to Growth: The 30-Year Update. In: OECD Observer (Issues 290–291). Earthscan
Melles G (2019). Participatory co-design for sustainable development. In: Leal Filho W, Azul A, Brandli L, Lange Salvia A, Wall T (eds) Partnerships for the Goals. Encyclopedia of the UN Sustainable Development Goals. Springer, Cham. https://doi.org/10.1007/978-3-319-71067-9_47-1
Melles G (2021). Figuring the transition from circular economy to circular society in Australia. Sustainability (Switzerland), 13(19). https://doi.org/10.3390/su131910601
Melles, G., Kuys, B., Kapoor, A., Rajanayagam, J., Thomas, J., & Mahalingam, A. (2015). Designing Technology, Services and Systems for Social Impact in the Developing World: Strong Sustainability Required. In A. Chakrabarti (Ed.), *Smart Innovation, Systems and Technologies* (Vol. 35, pp. 89–97). Springer India. https://doi.org/10.1007/978-81-322-2229-3_8
Melles G, Vere ID, Misic V, de Vere I, Misic V, Vere ID, Misic V, de Vere I, Misic V (2011) Socially responsible design: thinking beyond the triple bottom line to socially responsive and sustainable product design. CoDesign 7(3–4):37–41. https://doi.org/10.1080/15710882.2011.630473
Melles, G., Wölfel, C., Krzywinski, J., & Opeskin, L. (2022). Expert and Diffuse Design of a Sustainable Circular Economy in Two German Circular Roadmap Projects. *Social Sciences*, 11(9). https://doi.org/10.3390/socsci11090408
Merton R (1968) Social theory and social structure. The Free Press, New York
Millar, N., McLaughlin, E., & Börger, T. (2019). The Circular Economy: Swings and Roundabouts? *Ecological Economics*, *158*(April 2018), 11–19. https://doi.org/10.1016/j.ecolecon.2018.12.012
Morató T, González-Irusta JM, Domínguez-Carrió C, et al (2019) Climate-induced changes in the suitable habitat of cold-water corals and commercially important deep-sea fishes in the North Atlantic. Pangaea. https://doi.org/10.1594/PANGAEA.910319
Moreno M, De los Rios C, Rowe Z, Charnley F (2016) A conceptual framework for circular design. Sustainability (switzerland) 8:937.https://doi.org/10.3390/su8090937
Nelles M, Grünes J, Morscheck G (2016) Waste Management in Germany – Development to a Sustainable Circular Economy? Procedia Environ Sci 35:6–14. https://doi.org/10.1016/j.proenv.2016.07.001
Neumayer, E. (2003). Weak versus strong sustainability: Exploring the limits of two opposing paradigms, second edition. In *Weak versus Strong Sustainability: Exploring the Limits of Two*

Opposing Paradigms, Third Edition. Edward Elgar Publishing. https://doi.org/10.4337/9781849805438

Nilsson M, Griggs D, Visbeck M (2016) Policy: Map the interactions between Sustainable Development Goals. Nature 534(7607):320–322. https://doi.org/10.1038/534320a

Oliveira M, Miguel M, van Langen SK, et al (2021) Circular economy and the transition to a sustainable society: integrated assessment methods for a new paradigm. Circ Econ Sustain 1:99–113. https://doi.org/10.1007/s43615-021-00019-y

Olson RL (1995) Sustainability as a Social Vision. J Soc Issues 51(4):15–35. https://doi.org/10.1111/j.1540-4560.1995.tb01346.x

Papanek, V. (1971). *Design for the Real World: Human Ecology and Social Change* (2nd ed.). Bantam Books. http://www.amazon.com/Design-Real-World-Ecology-Social/dp/0897331532

Paredis, E. (2011). Sustainability transitions and the nature of technology. In *Foundations of Science* (Vol. 16, Issues 2–3). https://doi.org/10.1007/s10699-010-9197-4

Pepper D (1999) Ecological modernisation or the "Ideal model" of sustainable development? Questions prompted at Europe's periphery. Env Polit 8:1–34. https://doi.org/10.1080/09644019908414492

Platform for Accelerating the Circular Economy (PACE). (2020). *The Circularity Gap Report 2020: When circularity goes from bad to worse: The power of countries to change the game*. https://www.circularity-gap.world/

Ralph Boch, Gallen, J., & Hempel, N. (2020). *Wege zu einer Circular Society: Potenziale des Social Design für gesellschaftliche Transformation*. https://socialdesign.de/wp-content/uploads/2020/04/200420_HSS_Paper_CircularSociety_online.pdf

Raworth, K. (2017). *Doughnut Economics: Seven Ways to Think Like a 21st Century Economics*. Random House Business Books.

Raworth, K. (2019). Change the Goal: from GDP to the Doughnut. In *Doughnut Economicss* (pp. 27–50). Chelsea Green Pulbishing.

Rockström, J., & Steffen, W. (2009). Planetary boundaries: exploring the safe operating space for humanity. *Ecology and Society*, *14*(2), 32. http://www.ecologyandsociety.org/vol14/iss2/art32/main.html

Rofifah, D. (2020, March 11). *MITTEILUNG DER KOMMISSION AN DAS EUROPÄISCHE PARLAMENT, DEN RAT, DEN EUROPÄISCHEN WIRTSCHAFTS- UND SOZIALAUSSCHUSS UND DEN AUSSCHUSS DER REGIONEN: Ein neuer Aktionsplan für die Kreislaufwirtschaft Für ein saubereres und wettbewerbsfähigeres Europa*. 12–26.

Rudolph, C. (2018). Geschäftsmodell Circular Economy: Gegenwart und Zukunft der (erweiterten) Kreislaufwirtschaft. In P. Bungard (Ed.), *CSR und Geschäftsmodelle, Management-Reihe Corporate Social Responsibility* (pp. 123–137). Springer Verlag GmbH Deutschland. https://doi.org/10.1007/978-3-662-52882-2_7

Saidani, M., Yannou, B., Leroy, Y., & Cluzel, F. (2017). How to assess product performance in the circular economy? Proposed requirements for the design of a circularity measurement framework. *Recycling*, *2*(1). https://doi.org/10.3390/recycling2010006

Sanders EB-N, Stappers PJ (2008) Co-creation and the new landscapes of design. CoDesign 4(1):5–18. https://doi.org/10.1080/15710880701875068

Schandl, H., King, S., Walton, A., Kaksonen, A., Tapsuwan, S., & Baynes, T. (2021). *Circular economy roadmap for plastics, glass, paper and tyres -pathways for unlocking future growth opportunities for Australia* (Issue January). CSIRO Publishing.

Simon HA (1996) The Sciences of the Artificial (Third). MIT Press

Smith A (2004) Alternative technology niches and sustainable development. Innovation 6(2):220–235. https://doi.org/10.5172/impp.2004.6.2.220

Soergel B, Kriegler E, Weindl I, et al (2021) A sustainable development pathway for climate action within the UN 2030 Agenda. Nat Clim Chang 11:656–664. https://doi.org/10.1038/s41558-021-01098-3

Stahel, W. R. (2010). The Performance Economy. In *The Performance Economy* (2nd ed.). Palgrave Macmillan UK. https://doi.org/10.1057/9780230288843

Stahel, W. R. (2020). History of the Circular Economy. The Historic Development of Circularity and the Circular Economy. In *The Circular Economy in the European Union* (pp. 7–19). Springer International Publishing. https://doi.org/10.1007/978-3-030-50239-3_2
Stegeman H (2015) From circular materials cycles to a circular macroeconomy with scenario's for the Netherlands. Rabobank: Potential Circ Econ, 1–31. https://economics.rabobank.com/publications/2015/july/the-potential-of-the-circular-economy/. Accessed 5 Apr 2023
Sumter D, de Koning J, Bakker C, Balkenende R (2020) Circular Economy Competencies for Design. Sustainability (Switzerland) 12(4):1–16. https://doi.org/10.3390/su12041561
Velenturf APM, Archer SA, Gomes HI, Christgen B, Lag-Brotons AJ, Purnell P (2019) Circular economy and the matter of integrated resources. Sci Total Environ 689:963–969. https://doi.org/10.1016/j.scitotenv.2019.06.449
Velenturf APM, Purnell P (2021) Principles for a sustainable circular economy. Sustain Prod Consum 27:1437–1457. https://doi.org/10.1016/j.spc.2021.02.018
Wallner TS, Snel S, Magnier L, Mugge R (2022) Contaminated by Its prior use: strategies to design and market refurbished personal care products. Circ Econ Sust. https://doi.org/10.1007/s43615-022-00197-3
Weber T, Stuchtey M (2019) Pathways towards a German circular economy. lessons from european strategies preliminary study. www.circular-economy-initiative.de. Accessed 5 Apr 2023
Williams CC, Millington AC (2004) The diverse and contested meanings of sustainable development. Geogr J 170(2):99–104. https://doi.org/10.1111/j.0016-7398.2004.00111.x
Wilts H, Fink P (2016) Deutschland auf dem Weg in die Kreislaufwirtschaft. MÜLL Und ABFALL 15(6):1–4. https://doi.org/10.37307/j.1863-9763.2016.06.03
World Economic Forum (2020) Raising Ambitions: a new roadmpa for the automotive circular economy (Issue December).https://www.accenture.com/_acnmedia/PDF-146/Accenture-and-WEF-Raising-Ambitions-PoV.pdf#zoom=50. Accessed 5 Apr 2023

Beyond Eco-Design Towards Designing Sustainable Circular Production-Consumption Systems

2

Ursula Tischner

1 Resource Efficiency, Circularity, and Consumption

What are the challenges humanity is facing today regarding raw material consumption and ecosystem services? Several different research groups and approaches come to similar conclusions.

1.1 Resource Efficiency Research

In the early 1990s, Friedrich Schmidt-Bleek and his research group at the Wuppertal Institute for Climate, Environment and Energy proposed that Western industrialized countries have to dematerialize their production-consumption systems by a factor of 10 (Schmidt-Bleek and Klüting 1994). This group also came up with a definition of eco-intelligent consumption as consumption that is not based on material possessions but the resource-efficient use of goods (i.e. products, infrastructure, services). Eco-efficient consumption would then mean, selecting in each consumption decision the offer that provides the functions needed to fulfill the consumer's needs with the lowest possible use of materials and energy (Schmidt-Bleek et al. 1997).

Another member of the materials flow group at the Wuppertal Institute, Joachim Spangenberg, developed the Environmental Space concept, first presented in a report for Friends of the Earth Europe in 1995 (Spangenberg 1995). The

U. Tischner (✉)
econcept–Agency for Sustainable Design, Cologne, Germany
e-mail: u.tischner@econcept.org

Wilhelm Büchner Mobile University of Technology, Darmstadt, Germany

© The Author(s), under exclusive license to Springer Nature Singapore Pte Ltd. 2024
G. B. Melles and C. Wölfel (eds.), *Design for a Sustainable Circular Economy*,
Design Science and Innovation, https://doi.org/10.1007/978-981-99-7532-7_2

environmental space concept is based on Opschoor's initial definition of environmental space and identifies thresholds for resource consumption to secure non-deteriorating ecosystem services for future generations. Resource consumption should be reduced to a level at which the annual reduction of resources and their service potential can be compensated by newly discovered resources and efficiency gains in using them. In addition, Opschoor assumed equitable per capita consumption entitlements and concluded that a reduction of per capita consumption in industrialized countries by a factor of 8–10 was necessary (Opschoor 1987). Consequently, according to Spangenberg, the environmental space that every citizen of planet Earth can occupy is a space below the line of overconsumption (the 'ceiling' of the environmental space) but above a so-called poverty line or the line of dignity (the 'floor' of the environmental space). Thus, the lower line represents the minimum condition for **social sustainability**, while the upper line indicates **environmental unsustainability**. The environmental space between the two lines is a zone for free choice of consumption patterns. Comparable to monetary income, consumers have a budget for ecosystem services that they can spend on those goods and activities that are most important to them (Spangenberg 2014) (see Fig. 1).

Many years later, a similar model was developed by Raworth (2017) in her Doughnut Economics concept based on the planetary boundaries method developed by Johan Rockström and his team at Stockholm Resilience Centre (see below). Raworth (2017) in her concept also describes—like Spangenberg—that there is an upper limit for consumption (ecological ceiling) which is defined by the limits of our planet and that there is a minimum level of consumption (social foundation), i.e., everybody on this planet should be able to fulfill their basic needs like access to water, food or housing. In between these two circles is the safe and just space for humanity to live, according to Raworth (2017).

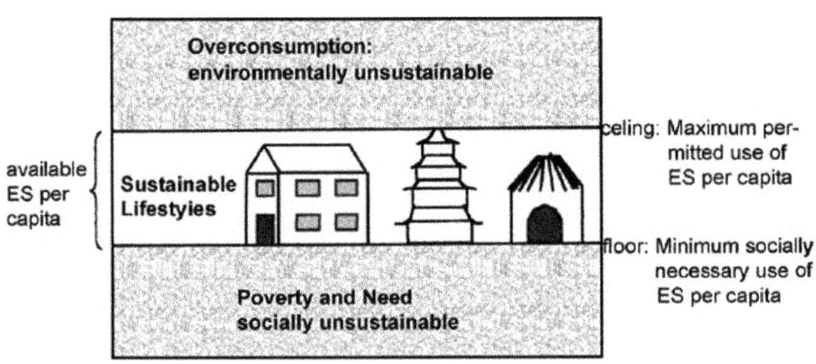

Fig. 1 The environmental space model developed at the Wuppertal Institute for Climate Environment and Energy, acc. to Spangenberg (1995)

1.2 The Circular Economy Concept

The idea of a Circular Economy is not new. In 1976, the US Congress passed the Resource Conservation and Recovery Act (US Congress 1976) to promote waste prevention, recycling, and resource conservation. Thus, the 3 Rs (Reduce, Reuse, and Recycle) were created as a slogan to communicate the idea to the population. Germany passed a circular economy and waste law in 1996, which was revised and updated in 2012 (KrW-/AbfG 2012). Both laws were developed within the framework of waste prevention approaches, i.e., they approach the issue from the waste perspective.

Re-think approaches that integrate circular economy strategies into the creation of products and business models were developed by Walter Stahel of the Institute for Product Life Cycle Research in Switzerland as early as the 1980s (Stahel 1991). He coined the term "performance economy" (Stahel 2010). At the end of the 1990s, the cradle-to-cradle design principle was formulated by the German chemist Michael Braungart and the American architect William McDonough. They suggested designing products in such a way that materials can be reused and recycled in natural or technical cycles, thus turning residual materials into nutrients (McDonough and Braungart 2002). Later on, the Circular Economy approach was widely publicized and promoted by Stahel and MacArthur (2019), the Ellen MacArthur Foundation.

Like the Resource Efficiency approach, the Circular Economy aims to decrease the consumption of virgin raw materials and to increase the intensive use of resources once they have been introduced to the technosphere, which also results in a reduction of waste. The means suggested by the Circular Economy are closing loops in the technosphere or the biosphere, especially by extending the useful lifespan of products through longevity and repairability, and by re-using as much as possible products, components, and materials on the highest value level in the technical systems. Alternatively, non-toxic bio-based materials can be cycled back to nature as food to natural systems. All of this should be based on using energy from renewable sources and supported by new circular business models. Digitalization can act as an enabler for these circular systems (Circular Economy Initiative Deutschland 2020).

1.3 The Ecological Footprint Concept

Another framework that pays regard to the need to cut global resource use is the Ecological Footprint concept by Mathis Wackernagel and the Global Footprint Network team (Rockström et al. 2009). They seek to capture the ecosystem services planet Earth provides and compare them with what people on this planet consume per country and overall. Some of the results of this type of calculation are that

- If everyone wanted to live like the average North American, this would require six planets,
- If everyone wanted to live like an average German, we would need four planets,
- Earth Overshoot Day, when humanity has consumed all the ecosystem services that the planet can renew within one year, takes place earlier each year. In 2022, it was already the 28th of July. To use an economic analogy: from that day on, humanity does not live from the interest but consumes the assets (the substance) of planet Earth (https://www.overshootday.org).

1.4 The Planetary Boundaries Research

Johan Rockström and his team at Stockholm Resilience Centre defined a framework of boundaries of ecosystem services that our planet offers. They described nine so-called Planetary Boundaries (Rockström et al. 2009). According to their research, humankind has already pushed climate change, biodiversity loss, shifts in nutrient cycles (nitrogen and phosphorus, very important for agriculture, i.e., food production), and land use beyond the boundaries into unprecedented territory. In addition, human societies have exceeded a planetary boundary related to environmental pollutants and other 'novel entities' including plastics (http://www.stockholmresilience.org) (see Fig. 2).

1.5 Overall Raw Material Reduction Targets

These different concepts all aim for the absolute reduction of global resource consumption. Industrialized countries (ICs) like Germany will have to decrease their resource consumption disproportionately in relation to the resource consumption of developing and emerging countries. Currently, reduction targets of a factor of 4 to a factor of 20 by 2030 or by 2050 are being discussed for ICs (e.g., Ressourcenkommission am Umweltbundesamt (KRU) 2014). The challenge is to secure prosperity, quality of life, and happiness for those who have access to them today, and to increase quality of life, prosperity, and happiness for those who lack it respectively, while at the same time reducing resource consumption overall. This level of reduction requires changes both in production and consumption and in the political framework conditions. It requires resource-efficient circular lifestyles as well as new economic models and radical technological and social innovations for more resource efficiency and circularity.

1.6 Role of Designers

This is a challenge for consumers and producers, politicians and academic researchers alike. But also designers have to ask themselves how they can design

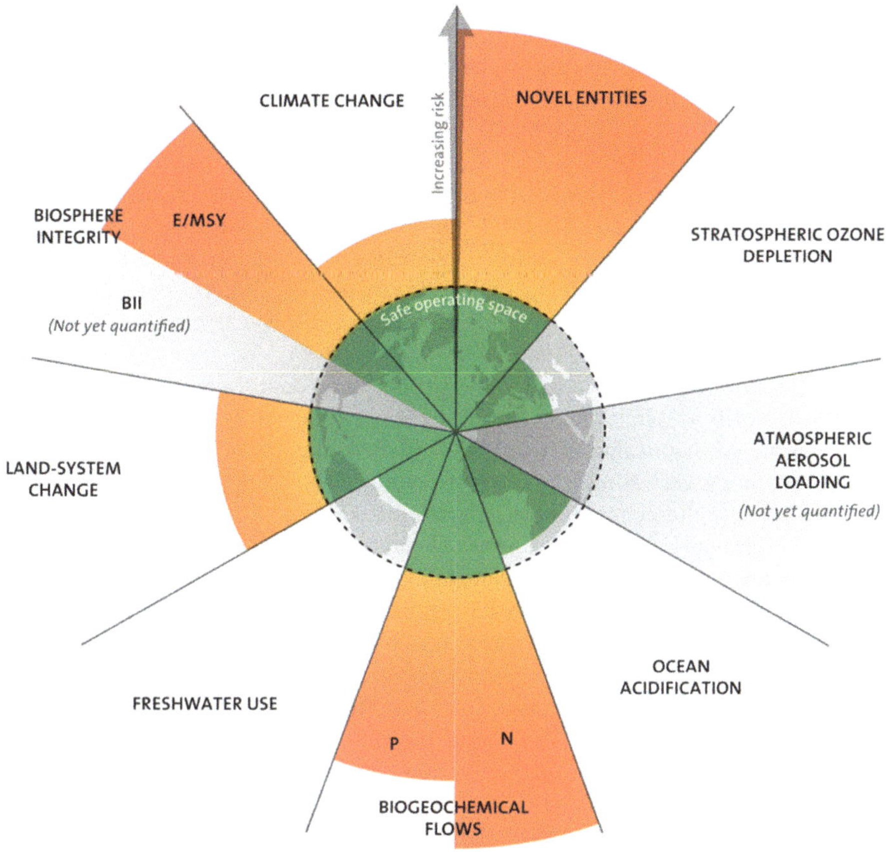

Fig. 2 Planetary Boundaries acc. to Stockholm Resilience Centre. *Source* Azote for Stockholm Resilience Centre

products, services, infrastructure, systems, communication, education, and social innovation that work towards this kind of 'absolute dematerialization' without losing or even gaining quality of life for as many people as possible. Altogether the quest is to encourage more sustainable, circular, resource-efficient lifestyles making them attractive to as many people as possible. Meanwhile, there is almost no product or service on the market that has not been touched by the hand or pen or computer of a designer, and it is the core competence of designers to make 'stuff' attractive. So why not put the skills of designers and other creatives on duty for the transformation towards sustainability instead of creating ever more problems in a throw-away consumer culture?

2 More Sustainable, Circular, Resource-Efficient Lifestyles

What are unsustainable lifestyles? These are lifestyles that cannot be sustained because they go against two fundamental limits: (a) the planet with its boundaries in terms of availability of land, fertile soil, resources, freshwater, i.e., everything that are so-called ecosystem services as described in the previous paragraph and (b) what is socially acceptable. If humankind is serious about equal human rights, then it is not fair that some citizens of this planet consume most resources and others live below the poverty line. Therefore, a sustainability evaluation of lifestyles has to take as a benchmark the more or less fair distribution of ecosystem services among all citizens of planet Earth. This distribution can be calculated on the basis of individuals or whole countries.

It might still be acceptable that some people in a country consume more when others voluntarily consume less. There might even be a trading scheme possible, where people who consume less can 'sell' parts of their share to people who like to consume more, very similar to Carbon Trading Schemes. It also makes sense to take into account the geography and climate of where people live, e.g., very cold or very hot climates need more heating or air conditioning, and food is easier to grow in temperate climates. These calculations of how much environmental space is available for each person in a specific location on planet Earth still are refined and values are negotiated. In an ideal world, however, every person living on this planet would have the equal right to access ecosystem services and each person would not consume more of its share than is available while keeping the planet in a healthy state.

2.1 Healthy Limits to Personal Consumption

The findings of the research groups described above are summarized as follows:

- In the logic of the ecological footprint, a sustainable lifestyle would mean, not consuming more than one planet.
- In the logic of the environmental space concept, this would mean staying above the line of social dignity but below the overconsumption line.
- In terms of resource efficiency, that would mean that people living in industrialized countries would need to reduce their resource consumption (including energy) on average at least by a factor of 4–10.

A closer look at consumption domains shows that the most impactful consumption activities with the highest potential for improvements in Europe have already been identified (European Environment Agency 2007). About 80% of all environmental impacts of European citizens are caused by three consumption domains:

- Housing and especially energy consumption in homes,
- Mobility and transportation,

- Food and agriculture.

Very likely this is similar for most industrialized countries.

2.2 How to Calculate One's Own Footprint

Some simple online tools exist to calculate one's personal environmental impact, such as:

- For resource footprints (or ecological rucksack) the resource calculator of Wuppertal Institute (https://www.ressourcen-rechner.de/?lang=en);
- For ecological footprints the ecological footprint calculator of Global Footprint Network (https://www.footprintcalculator.org/home/en);
- And for carbon footprints the carbon footprint calculator from Carbon Footprint Ltd. (https://www.carbonfootprint.com/calculator.aspx).

The following table shows the individual footprints calculated with the tools above for a person living in Germany. This person eats hardly any meat and buys mainly seasonal, regional, and organic products. She has no car and travels mainly by public transport, bike, and foot but has to fly for quite a few trips. She lives alone in an apartment of 56 m^2 that is heated by district heating and has a green power contract. The person uses limited amounts of products consciously, often buys second-hand, and exchanges products only when they are broken (see Table 1).

While these results are not comparable, they lead to similar conclusions: Individual consumer behavior is only partly responsible for a person's environmental impact. The services provided in the country where a person lives are also decisive, which means consumers alone cannot be made responsible for the resource efficiency of a country. The infrastructure and public services among others have to be taken into account and improved as well.

In the case of the person above, one major environmental impact emerges in the mobility domain from flying per year. In the ecological footprint calculation, 70% of the person's impact is made up of her carbon footprint, and four of the seven global hectares the person consumes come from flying. That is her 'big issue'.

Table 1 Results of individual environmental impact calculators for the same person, compared (own calculation)

Measuring system	Results
Resource footprint Target state would be 17 tons	19.1 tons are consumed by this person Compared to an average of 27 tons for all users of the calculator
Ecological footprint Target state would be 1 planet	4.2 earths would be needed, if everybody lived like this person Earth overshoot day of this person is 29th of March
Carbon footprint Target state would be 2 tons	CO_2 emissions of this person are 5.03 tons The average of people living in Germany is 8.89 tons

Without taking into account the flights she would only be 'responsible' for three global hectares or around two planets.

For other people that 'big issue' might be eating meat, driving a car every day, or buying and disposing of a lot of products and clothing, etc.

That means, as diverse as the consumption patterns are, as multifold are the options for improvement. There is no such thing as 'the one sustainable lifestyle', but there are many different ways in the various consumption domains to live more sustainably. This relation is especially true across different cultures and consumer groups. *Diversity and freedom of choice are key—also for the acceptance of such more sustainable lifestyles.* That is why the environmental space concept gives an upper and lower limit for the consumption of resources, energy, and land, but does not prescribe how citizens have to live. Every person still has many individual choices on how to'spend' their share of the ecosystem services.

2.3 How to Reduce One's Footprint

Nevertheless, one can formulate a few generic measures as priorities that enable more sustainable living for a relatively homogenous consumer culture, such as the one in industrialized countries, as follows (see Table 2).

If these are some of the most important measures to reduce consumer's individual environmental impact and resource consumption in industrialized countries, then the next question is, how can consumers be attracted to implementing these and similar measures? This will likely be most successful if the behavior changes will contribute to their well-being and happiness.

3 Happiness and Wellbeing

The happiness and well-being quest has become an industry—from the first books about 'simplifying your life' to Marie Kondo (https://konmari.com) helping us declutter our wardrobes and our lives, from uncountable magazines, websites, and coaching services to educational programs on how to live happily. Indeed, our lives in the modern digital world have become stuffed and filled with digital gadgets and information overload all around. Work and consumption pile up in our waking hours and haunt us in our nocturnal dreams. Today, time has become a scarce resource, and it is almost a social obligation to search for one's own happiness.

3.1 Unhappy Consumers

The multi-optional consumerist society is deeply unhappy and unhealthy. Stress and burnout rates are at a high level and still increasing. According to the AOK, a German public health insurance fund insuring around 1/3 of the German population, an average of one burnout case per 1,000 AOK members was

Table 2 Consumption domains and most important sustainability measures. Own compilation based on findings from the European SusProNet (cf. Tukker and Tischner 2006) and SCORE (cf. Tischner et al. 2010) research projects

Consumption domain	Measure
Housing/energy consumption in homes	Reduce the space you occupy for housing or increase the number of people sharing the space
	Take care of good insulation and construction of your home so that heating/cooling needs are reduced
	Select a green energy provider or produce your own green energy
	Reduce the number of products that consume power in your home, use them efficiently, and prefer energy-efficient devices
Mobility/transport	Reduce the need for mobility/transport, e.g., by living close to where you work, buying local, going on holidays close to where you live
	Prefer efficient mobility means: Walking/biking before bus, train, and car
	Prefer efficient transportation means: Bicycle courier before ship, train, and truck
	Avoid flying and transportation by air
Food/agriculture	Eat no or less meat
	Consume no or less dairy products
	Prefer regional, seasonal, organic, unpacked, unprocessed produce
	Eliminate/reduce food waste

diagnosed in Germany in 2005; in 2017, there were already 5.5 cases of incapacity for work per 1,000 members. According to the AOK, the number of diseases due to burnout diagnoses has more than quadrupled in the last ten years (https://de.statista.com/statistik/daten/studie/239672/umfrage/berufsgruppen-mit-den-meisten-fehltagen-durch-burn-out-erkrankungen/). Burn-out is associated with a high burden in work and private life. Similar increases can be detected in other countries (Abramson 2022).

3.2 Aspects of a 'Good' Life

When asking what makes us happy, again, it is easier to understand what makes us unhappy. Interesting anecdotal findings in this context, come from the Australian palliative nurse Bronnie Ware talking to her patients, who knew that their lives would soon end, about what they wished they had done differently. Common themes surfaced again and again. The most common five were (Ware 2012):

1. I wish I had the courage to live a life true to myself,
2. not the life others expected of me.
3. I wish I hadn't worked so hard (especially male patients).
4. I wish I had the courage to express my feelings.
5. I wish I had stayed in touch with my friends.
6. I wish that I had let myself be happier.

It seems that none of the dying persons Ware nursed wanted more money, a better job, a faster car, or more beautiful clothing, etc. All of the regrets of the dying are about immaterial issues such as self-actualization, developing one's own talents, social connection to others, and living a life true to oneself. In Maslow's hierarchy of needs, these are the needs of self-actualization and transcendence, self-esteem, and the social needs of love and belonging, not the physiological or safety needs (Maslow 1943).

Beyond the anecdotal, research about happiness and well-being was carried out early on by economist Richard Layard. He detected that the happiness of people in industrialized countries, e.g., the US, is not automatically correlated to higher income. Between 1956 and 1996, the percentage of very happy people in the US declined despite a constant increase in GDP per capita (Layard 2005).

Layard is also co-editor of the World Happiness Reports of the UN. These reports collect and discuss findings about the state of happiness in the world. In the 2019 issue, for instance, the connection between the use of digital media and well-being has been analyzed and the authors concluded that longer screen time is correlated to lower well-being of adolescents. Vice versa reducing the use of digital devices and media increases happiness. These might be rather indirect effects: screen time takes away time for other activities that contribute to well-being. The report refers to many scientific studies (Hartgerink et al. 2015; Lieberman 2014; Zhai et al. 2015) concluding that deprivation of social interaction and lack of sleep are clear risk factors for unhappiness and low well-being in adults (Helliwell et al. 2019).

Similar conclusions come from the Greater Good Science Center, based at the University of California Berkeley, which started in 2001 to collect academic studies from psychology, sociology, and neuroscience around the well-being of people with the aim to offer the gathered knowledge and education on how to build a thriving, resilient, and compassionate society to the public. The center promotes the concept that individual happiness is connected to strong social bonds, altruistic and cooperative behavior, mindfulness and compassion, and overall a meaningful life (https://ggsc.berkeley.edu/what_we_do/event/the_science_of_happiness).

To conclude, earning more money, above an income that allows us to fulfill our basic physical and safety needs, and spending it on more adorable products, entertaining us in front of computers and mobile phones, having a lot of friends on social media, and other seemingly satisfying activities might not deliver to us the happiness and well-being that marketing and advertising agencies and the companies work for to make us believe. It might rather be true what Annie Leonard says in the 'Story of Stuff' video clips: "*We are trashing the planet, we are trashing the people and we are not even having fun.*" (https://storyofstuff.org/movies/). Important first steps towards more sustainable lifestyles would be to educate consumers about these findings early on, e.g., in schools, and to promote more ethical advertisements.

As the core competence of designers is to make goods attractive for consumers/ users with the promise that buying and using the goods will make them happy or will be good for them, the next paragraph discusses the role of design and designers in this context.

4 System Design for Sustainability and Circularity

The traditional position of designers and design is in the middle between production and consumption. Designers work for companies that like to sell 'stuff' and design that 'stuff' (product-, industrial, user experience design, also software design) and the communication (communication-, graphics-, media design) and other services (service design) around it so that the targeted consumer or user groups like and buy the offer. Thus, designers normally have two types of clients: the producers and users/consumers. But because they are paid by the producers, their interest is a little more important to the designers. Environment had no agency in the design process until the 1970s when it became clear that the throw-away society creates considerable damage to the natural environment.

4.1 From Eco-Design to System Design for Sustainability

Industrial designers like Victor Papanek described early on that the design as taught and practiced was creating more problems than solutions (Papanek 1985). Then Green or **Eco-Design** (Tischner et al. 2000) was defined as a different design discipline taking into account environmental issues alongside other common aspects like function, aesthetics (Papanek 1985, 1995), and price, and covering the whole life cycle and product system of a product. The circularity of products, components, and materials was already part of the Eco-Design method and strategies (Tischner and Moser 2015).

Because of constantly increasing global consumption and the rebound effects that counteracted the improved eco-efficiency of the products (more products are used, or the efficient products are used in very in-efficient ways), it became clear that the whole system of production and use of products would have to be re-designed; not just the products alone. Thus, the approach of **Product-Service System design (PSS)** for Sustainability was developed (Tukker and Tischner 2006). Here the focus is on fulfilling the needs with the most efficient combination of products and services moving as much as possible towards immaterial services rather than selling a lot of material products, e.g., as is the aim of the sharing or servicizing economy. The PSS concept was the transition to an even more systemic approach to designing or influencing overall production and consumption systems such as mobility, food and agriculture or energy production and consumption, that was also supported by the United Nations under the term **Sustainable Consumption Production (SCP)** system design or **System Innovation for Sustainability** (Tischner 2008a, b).

4.2 More Radical System Design Approaches

Although these more complex systemic design approaches are still far away from being regular design practice, there are more and more designers suggesting similar radically different approaches to design, because these are needed to tackle the big and complex challenges for humanity as described in the previous paragraphs. Upcoming terms are eco-centric or humanity-centered design instead of human-centered design. Other examples are:

- **Transformation Design**, first suggested by the Design Council in London around 2003 and applied in the RED projects, is a human-centered interdisciplinary process that aims at creating desirable and sustainable changes in behavior and form of individuals, systems, and organizations. The process is multi-staged, iterative, and applied to big, complex issues—often social issues. The challenges are holistically examined, and then new small-scale systems including objects, services, interactions, and experiences are prototyped that support people and organizations in achieving the desired change. Successful prototypes can then be scaled. The RED projects have resulted in the creation of new roles, new organizations, new systems, and new policies (Burns et al. 2006).
- **Speculative Design**, a term suggested by Anthony Dunne and Fiona Raby is a discursive, research-oriented, experimental design approach based on critical thinking and dialog that aims at including the public in the rethinking and envisioning of and dialog on new technological realities and new social relations. Dunne and Raby had explored the potential of new technologies for the future issues of our time under the label 'Critical Design' and then moved to Speculative Design as a method to initiate discussions—not to offer concrete solutions that can be implemented directly. However, a successful Speculative Design project is necessarily connected to the research of a social context and is fundamentally directed towards individual needs and desires (Dunne and Raby 2013). Its results are often imaginations of desired futures and visions of possible scenarios. Speculative Design is transdisciplinary or even post-disciplinary in nature which means it relies on interactions between various disciplines.
- **Transition Design**, as promoted by Terry Irwin and her colleagues at the Carnegie Mellon University acknowledges that societal transitions are happening and needed to reach more sustainable futures, and argues that design has a key role to play in these transitions. The interconnected fields of social, economic, political, and natural systems are taken into account to suggest a

rethinking of entire lifestyles with the aim of harmonizing them with the natural environment and making them more place-based, convivial, and participatory, yet global in their awareness and exchange of information and technology. Basic human needs shall be satisfied locally, within economies that exist to meet those needs, as an opposite to the mainstream dominating economic growth and profit-maximizing paradigm. Transition designers draw on knowledge and wisdom from the past to conceive solutions in the present with future generations in mind. The tools to do so come from complex systems theory, transition, and change management, as well as philosophy, psychology, social science, and anthropology among others (Irwin et al. 2015).

4.3 Common Elements of System Design for Sustainability

All these new and radically different approaches to design have common elements that are also intrinsic to **System Design for Sustainability**. These are mainly the following characteristics:

- System perspective, beyond a product, or a service, or communication, or a user experience design. System design analyses complex systems, and identifies problems—even wicked problems (Rittel and Webber 1973) and opportunity spaces for many different actors (users, producers, service providers, etc.) to act more sustainably.
- Multidisciplinary, Cross-disciplinary or Transdisciplinary, across the design disciplines, of designers with many other disciplines, and between other disciplines.
- Multi-stakeholder approaches and Co-Creation: Diverse groups of stakeholders are involved. The approach is often participatory from the research phase over the design to the implementation of the results. This approach requires empathy and collaboration/co-creation skills on the side of designers, as well as high-level facilitation and organizational skills.
- They aim at radical new solutions and disruptive changes in existing systems to create a more sustainable society focusing on all three pillars of sustainability as defined by the UN—environment, social, and economic (World Commission on Environment and Development 1987).
- They acknowledge that this change is not only about technological innovation, but also social and organizational innovation might be equally or even more important, ultimately aiming at encouraging more sustainable ways of living.
- Often new start-ups, services, business ideas, and business models emerge in the context of system design projects. Here, the business model canvas (Osterwalder et al. 2010) and work done by Oliver Gassmann and the team at the University of St. Gallen on business model innovation patterns is helpful (Gassmann et al. no date). Even circular business model innovation guides are available (e.g. https://bmilab.com/topics/circular-economy).

5 From Circular Design to Design of Sustainable Circular Systems

So far, Circular Design is often understood as focusing on the circular design of products by using circular design strategies like 'the Rs' of circular design: Reduce, Repair, Reuse, Remanufacture, and Recycle. This is a good first step. In addition, 'the Ds' of Circular Design should be considered as well; we should design so that it is possible to De-polymerize polymers, De-alloy metals, De-laminate composites, De-vulcanize rubber, De-coat materials, and De-construct high-rise buildings and major infrastructure (Stahel and MacArthur 2019).

However, this is still not enough to guarantee sustainability. Every circular process still needs energy, transportation, collection, sorting, remanufacturing, recycling technologies, etc. Many of these technologies are still lacking. In addition, according to the 2nd law of thermodynamics, simply put, these transformation processes increase the entropy in the system. As long as the share of renewable energies is too low—according to Eurostat, it was around 22% on average in the European Union in the year 2020 (https://ec.europa.eu/eurostat/cache/infographs/energy/bloc-4c.html), the energy needed for recovery and recycling processes still increases climate change. Thus, all energy sources used in a circular economy should be renewable.

Furthermore, some processes can simply not be reversed, e.g., dissipative losses of materials such as tire abrasion, aerosols or microplastics. Thus, all of these dissipative materials should be non-toxic and digestible for nature. Generally, with more and more circulation of materials, toxic and hazardous substances contained in the materials will accumulate in the system. Thus, the use of toxic and hazardous substances should be eliminated or reduced to a minimum.

In addition, social aspects of the circular economy, such as what kind of jobs are lost and created by circular systems and who will do the work, are often neglected.

And finally, recycling of materials cannot meet the increasing demand, if resource consumption continues to rise globally as it has been the case in the past (International Resource Panel 2019). For instance, in the year 2019, over 14 million tons of plastics have been processed in Germany. The share of recyclates was almost 2 million tons, but only 430,000 tons post-consumer recyclates substituted virgin plastics (Conversio Study 2020). Therefore, carefree overconsumption has to be stopped, especially in industrialized countries, and the upper and middle classes in developing and emerging countries need to change their behavior towards more sufficiency.

All of these arguments point towards the conclusion that the circular economy and circular design alone will not make our production-consumption systems more sustainable, if we continue or even increase current consumption levels. What is needed is a more holistic design of sustainable production-consumption systems that includes circularity but goes beyond that; starting with the question of what is really needed by consumers and society to lead a good life and how we can deliver that in the most sustainable way. It continues with considerations about new (circular) business models and social as well as technical innovations needed, ultimately also questioning the underlying economic system.

5.1 Activities in a System Design for Sustainability Methodology

A design methodology to design production-consumption systems for increased sustainability should consist of the steps and activities described below. In all of these activities, users/consumers and stakeholders including political actors should be involved as much as possible (participatory/Co-Design). Circularity in the systems is an important element, but even more important are the radical approaches of Re-thinking systems and Reducing consumption:

Step 1: Holistic system analysis	Thoroughly analyze the existing situation to identify major problems/sustainability issues, such as where is the largest resource consumption, what are social issues, and where is the biggest improvement potential, using, for instance, the sustainability SWOT analysis, identify windows of opportunities for improvement
Step 2: Scenarios, backcasting, and roadmap	Scenario building and backcasting: for those areas you have identified, imagine desirable futures and how well-being can increase while resource consumption and environmental impacts can decrease for as many stakeholders involved in the system as possible, start where the system is now and develop potential road maps (steps over a defined period of time) to reach the most desirable and sustainable futures
Step 3: Selection of focus and ideation	Focus on the most promising areas and ideate design interventions (products, services, social innovations, education, communication, bottom-up initiatives, and other activities) that can help to move from one step to the next step in the roadmap towards the desired sustainable future
Step 4: Selection of solutions and detailing business/ operational models, networks, etc.	Evaluate ideas according to sustainability improvements, e.g., using simplified sustainability screening tools or even Screening LCAs, and identify the most promising sequence of steps. Formulate business models and financing models for the design interventions (including innovative ones such as crowdfunding), identify the partners needed to implement, and prepare implementation

(continued)

(continued)

Step 5: Implementation	Implement the steps starting with the first. Evaluate progress and learn, and adapt accordingly. If some measures fail, redesign the steps as needed. Start with small-scale application/test in niches, if sensible, and then scale up or multiply

6 How to Make More Sustainable Production-Consumption Desirable

There will only be a social change towards sustainability, if we change personally, if people change in their private lives and at work. One can call for institutions, politicians, and companies, but who is the politician, and who is the entrepreneur? They are all people. In other words, what is needed is a change in the mindsets and behavior of people. If people change their behavior, then organizations change too. However, change is difficult, change creates fear, people may be a bit lazy and uncomfortable in their situation, afraid of the unknown, and hesitant to break new ground. Many studies have shown that people feel secure in their routines. According to behavioral economics, humans like to stick to known routines, because that needs the lowest investment in terms of time and effort for decision-making. Thus, about 80% of everyday behavior is 'routine' (Tischner et al. 2010). People always drive the same way to work, buy the same things in the supermarket, go to well-known restaurants, and so on. To change the learned and routinised behavior is associated with effort and the outcome is uncertain. For example, we don't know exactly whether the food in a new restaurant is as good as in the usual one. The same applies to new products that are unfamiliar, require us to behave differently or might be somewhat unusual. There are adventurous people that always search for new exciting experiences, the trendsetters, but in Germany, they make up only around 10% of the population (acc. to the SINUS Milieus Typology: https://www.sinus-institut.de/sinus-loesungen/sinus-milieus-deutschland/).

One tool to overcome procrastination is the so-called 'nudging', an often subtle, interesting, fun, and humorous way to encourage better consumer behavior, e.g., by design interventions (Thaler and Sunstein 2009). Edutainment, gamification, starting of communities where like-minded people encourage each other are other approaches, as well as the involvement of users/consumers in the design and production of the desired outcomes. The more consumers are also involved emotionally and actively and the more the new behavior feels good and generates positive results, the more likely they stay with it. Ultimately it needs a new narrative away from a consumer culture towards a well-being culture in society, where identity is not mainly connected to the stuff we own and the number of social media contacts we have.

According to behavioral psychotherapy learning new behaviors includes four elements (Kanfer and Schmelzer 2005):

(1) Awareness: one becomes aware that a problem exists,
(2) Motivation: one has (intrinsic or extrinsic) motivation to change one's habitual behavior,
(3) Opportunities: one then has the opportunity to try out a new behavior in a safe space, and finally
(4) Positive Reinforcement or Reward: When trying the new behavior, there should be positive reinforcement or reward, e.g., saving money, a satisfying experience, or admiration by peers, and so on. Thus, the new behavior proves successful and pleasant and can very likely be integrated into daily routines as a positive behavior pattern.

Designers can support this learning cycle towards sustainable lifestyles:

(1) They can communicate better and more appropriately to the target group (storytelling, edutainment, etc.) to create more awareness of sustainability issues and the connection to personal behavior—one of the core competencies of communication designers.
(2) They can try to strengthen extrinsic or intrinsic motivation, e.g., by showing role models and positive examples, or by making the abstract sustainability issues more emotional, explaining how things work in other cultures, how sustainability can look and feel nice, and how it can be fun, among others.
(3) Designers can create opportunities and possibilities for more sustainable behavior by designing new products, services, and systems (core competencies of product designers). They are involved in designing more sustainable infrastructure, products, services, and social innovations, and these are precisely the opportunities that can enable people to act more sustainably.
(4) And finally, to ensure that the new behavior becomes a positive experience, designers can help to organize positive feedback, e.g., through communities and peer groups, by saving money, just feeling better, and having a better conscience, among others.

An excellent collection of many different design interventions to encourage better consumer behavior, from very authoritarian ones to fun and gaming, can be found in the Design with Intent Cards by Dan Lockton (http://designwithintent.co.uk). When trying to influence consumer/user behavior there are fine lines between manipulation, seduction, and offering alternative opportunities as choices. Consumer choices are often not purely rational but involve quite a bit of emotion. Transparent, trustworthy, and truthful storytelling, offering honest education and information should be the first choice, but sometimes a bit of seduction is needed to change routinized behavior.

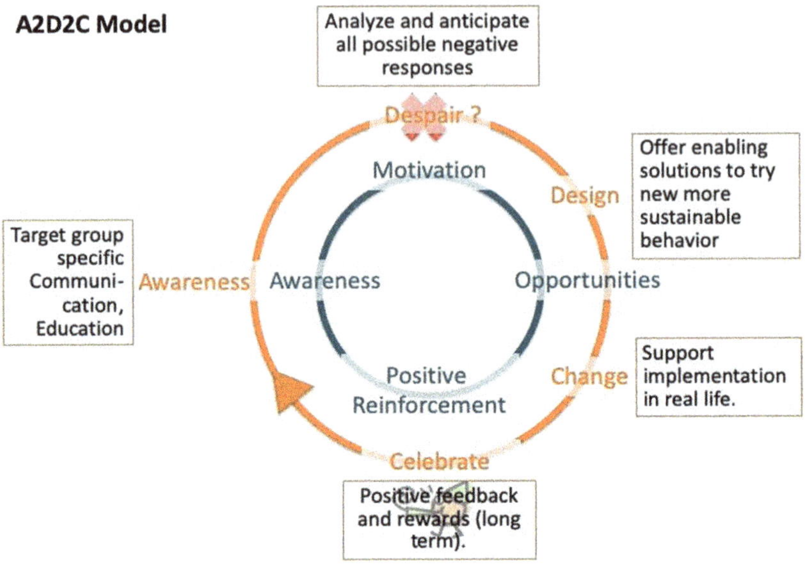

Fig. 3 Awareness, Despair, Design, Change, Celebrate: A2D2C Model. *Source* Stebbing and Tischner (2015)

Figure 3 shows the 'A2D2C Model' (Awareness, Despair, Design, Change, Celebrate), which has been developed by the author as a tool for designers illustrating the interrelations described above.

7 How to Implement System Design for Sustainability

System design for sustainability is an exciting and emerging field for designers and other creatives. For its implementation, an important question remains: Who commissions and pays the designers if they might not be the producer of a product anymore? Fortunately, there are also new financing and funding schemes emerging that can help to fund these kinds of activities. Crowdfunding, crowd donating, and crowd investing for instance (e.g., www.kickstarter.com, https://www.ecocrowd.de, https://www.leetchi.com, https://www.betterplace.org/de), where a larger group of people, especially those that are affected or sympathized with the issues at hand, are invited mainly via online platforms to collectively finance the project—with or without (donation) rewards for them (Tischner and Beste 2016). There might be institutions, other than companies, funding these kinds of projects from public organizations to foundations and NGOs. It is possible to apply for research funding, if the projects have some kind of research question to answer, as most of them have. And finally, there are indeed also companies and start-ups out there that drive and finance these kinds of more radical and more sustainable design and innovation projects, because they are more interested in long-term positive impact than in short-term profits.

Like everybody else in these systemic design projects for sustainability, designers and other creatives are challenged to move out of their comfort zones. They need to learn and apply new methods and ask and answer more fundamental and more complex questions. Hopefully, these methods and tools will be taught in schools and universities in the future—at the moment there are only very few design programs available where this is offered(according to the research and academic experience of the author and her work in the university accreditation organizations). Once involved in projects like this, designers can also radically increase their positive impact, and this might lead to more happiness—not just for stakeholders, but also for the creatives themselves.

References

Abramson A (1 Jan 2022) Burnout and stress are everywhere. In: 2022 trends-report, vol 53, no 1. American Psychological Association, p 72

Burns C, Cottam H, Vanstone C, Winhall J (2006) RED PAPER 02: transformation design. Design Council, London

Circular Economy Initiative Deutschland (ed) (2020) Circular business models: overcoming barriers, unleashing potentials, acatech/SYSTEMIQ, Munich/London

Conversio Study (2020) Material flow analysis plastics in Germany 2019. https://www.conversio-gmbh.com/res/News_Media/2020/Summary_Material_Flow_Analysis_Plastics_Germany_2019_EN.pdf

Dunne A, Raby F (2013) Speculative everything: design, fiction and social dreaming. MIT Press, Cambridge

European Environment Agency (2007) Environmental pressures from European consumption and production. EEA Publication TH-78-07-137-EN-D

Gassmann O, Frankenberger K, Csik M (no date) The St. Gallen business model navigator. University of St. Gallen, BMI Lab, St. Gallen. https://www.thegeniusworks.com/wp-content/uploads/2017/06/St-Gallen-Business-Model-Innovation-Paper.pdf. Accessed 21 Oct 2019

Hartgerink CHJ, van Beest I, Wicherts JM, Williams KD (2015) The ordinal effects of ostracism: a meta-analysis of 120 cyberball studies. PLoS One 10(5). https://journals.plos.org/plosone/article?id=10.1371/journal.pone.0127002. Accessed 21 Oct 2019

Helliwell JF, Layard R, Sachs JD (eds) (2019) World happiness report 2019. Sustainable Development Solutions Network, New York. https://s3.amazonaws.com/happiness-report/2019/WHR19.pdf. Accessed 21 Oct 2019

International Resource Panel (2019) Global Resources Outlook 2019: Natural resources for the future we want. Report of the international resource panel. United Nations Environment Programme, Nairobi

Irwin T et al (2015) Transition design 2015. School of Design, Carnegie Mellon University, Pittsburgh

Kanfer FH, Schmelzer D (2005) Wegweiser Verhaltenstherapie: Psychotherapie als Chance (2 Aufl). Springer, Berlin

Kreislaufwirtschaftsgesetz vom 24. Februar 2012 (BGBl. I S. 212), das zuletzt durch Artikel 2 Absatz 2 des Gesetzes vom 9. Dezember 2020 (BGBl. I S. 2873) geändert worden ist. https://www.gesetze-im-internet.de/krwg/KrWG.pdf

Layard R (2005) Happiness: lessons from a new Science. Allen Lane, London

Lieberman MD (2014) Social: why our brains are wired to connect. Broadway Books, New York

McDonough W, Braungart M (2002) Cradle to cradle: remaking the way we make things. North Point Press, New York

Maslow AH (1943) A theory of human motivation. Psychol Rev 50(4):370–96. https://citeseerx.ist.psu.edu/viewdoc/summary?doi=10.1.1.334.7586. Accessed 21 Oct 2019

Opschoor JB (1987) Duurzaamheid en verandering: over ecologische inpasbaarheid van economische ontwikkelinge [Sustainability and change: about the environmental suitability of economic developments]. Free University Press, Amsterdam

Osterwalder A, Pigneur Y, Clark T (2010) Business model generation: a handbook for visionaries, game changers, and challengers. Strategyzer series. Wiley, Hoboken

Papanek V (1985) Design for the real world: human ecology and social change. Academy Chicago Pub. Ltd., Chicago

Papanek V (1995) The green imperative: ecology and ethics in design and architecture. Thames & Hudson Ltd., London

Raworth K (2017) Doughnut economics: seven ways to think like a 21st-century economist. Random House

Ressourcenkommission am Umweltbundesamt (KRU) (Hrsg) (2014) Ressourcenleicht leben und wirtschaften, Standortbestimmung der Ressourcenkommission am Umweltbundesamt (KRU). Umweltbundesamt, Berlin/Dessau

Rittel HWJ, Webber MM (1973) Dilemmas in a general theory of planning. Policy Sci 4:155–169. http://urbanpolicy.net/wp-content/uploads/2012/11/Rittel+Webber_1973_PolicySciences4-2.pdf. Accessed 21 Oct 2019

Rockström J et al (2009) Planetary boundaries: exploring the safe operating space for humanity. Ecol Soc 14(2). Art 32

Schmidt-Bleek F, Klüting R (1994) Wieviel Umwelt braucht der Mensch?: MIPS Das Maß für ökologisches Wirtschaften. Birkhäuser, Berlin

Schmidt-Bleek F, Merten T, Tischner U (Hrsg) (1997) Ökointelligentes Produzieren und Konsumieren (Wuppertal Texte). Birkhäuser, Basel

Spangenberg JH (ed) (1995) Towards sustainable Europe. a study for friends of the earth Europe. Russel Press, Nottingham, UK

Spangenberg JH (2014) Institutional change for strong sustainable consumption: sustainable consumption and the degrowth economy. Sustain Sci Policy Pract 10(1):62–77. Open access at https://www.tandfonline.com/doi/pdf/10.1080/15487733.2014.11908125?needAccess=true. Accessed 21 Oct 2019

Stahel W (1991) Vertiefungsstudie zu Langlebigkeit und Materialrecycling im Bereich der Produkte. Stuttgart

Stahel W (2010) The performance economy, 2nd ed (1st ed, 2006). Palgrave-MacMillan, London

Stahel WR, MacArthur E (2019) The circular economy. A user's guide. Routledge, London UK

Stebbing P, Tischner U (2015) Changing paradigms. Designing for a sustainable future. Cumulus Association/Aalto School of Design, Helsinki. Accessible via https://www.cumulusassociation.org/changing-paradigms-designing-for-a-sustainable-future/. Accessed 21 Oct 2019

Thaler RH, Sunstein CR (2009) Nudge: improving decisions about health, wealth and happiness. Penguin, London

Tischner U et al (2000) How to do ecodesign? A guide on environmentally and economically sound design, form Vlg., Frankfurt

Tischner U (2008a) Chapter 1: introduction. In: Tukker A et al (eds) System innovation for sustainability 1: perspectives on radical changes to sustainable consumption and production. Greenleaf Publishing, Sheffield UK

Tischner U (2008b) Chapter 9. Design for (social) sustainability and radical change. In: Tukker A et al (eds) System innovation for sustainability 1: perspectives on radical changes to sustainable consumption and production. Greenleaf Publishing, Sheffield UK

Tischner U, Stø E, Unni K, Tukker A (2010) System innovation for sustainability 3, case studies in sustainable consumption and production–food and agriculture. Greenleaf Publishing, Sheffield UK

Tischner U, Moser H (2015) How to do Ecodesign, practical guide for Ecodesign–including toolbox, Umweltbundesamt available as E-Book: https://www.umweltbundesamt.de/publikationen/how-to-do-ecodesign, and I-Book: https://itunes.apple.com/de/book/how-to-do-ecodesign/id1135764180?mt=11. Both accessed 21 Oct 2019

Tischner U, Beste L (2016) State of the art of open innovation and design for sustainability. In: Matsumoto M, Masui K, Fukushige S, Kondoh S (eds) Sustainability through innovation in product life cycle design. Springer, Singapore, pp 705–717

Tukker A, Tischner U (2006) New business for old Europe, product-service development, competitiveness and sustainability. Greenleaf Publishing, Sheffield UK

US Congress (1976) Resource conservation and recovery act of 1976, Public Law 94-580 94th Congress, USA

Ware B (2012) The top five regrets of the dying: a life transformed by the dearly departing. Hay House Inc., Carlsbad

World Commission on Environment and Development (1987) Our common future. Oxford University Press, Oxford

Zhai L, Zhang H, Zhang D (2015) Sleep duration and depression among adults: A meta-analysis of prospective studies. In: Depression and anxiety, vol 32, pp 664–670

Internet Links

https://ellenmacarthurfoundation.org

https://environment.ec.europa.eu/topics/circular-economy_en

http://www.footprintnetwork.org/en/index.php/GFN/page/footprint_basics_overview/. Accessed 21 Oct 2019

https://www.overshootday.org. Accessed 21 Oct 2019

http://www.stockholmresilience.org. Accessed 21 Oct 2019

https://www.ressourcen-rechner.de. Accessed 21 Oct 2019

https://www.footprintcalculator.org. Accessed 21 Oct 2019

https://www.carbonfootprint.com/calculator.aspx. Accessed 21 Oct 2019

https://konmari.com

https://de.statista.com/statistik/daten/studie/239672/umfrage/berufsgruppen-mit-den-meisten-fehltagen-durch-burn-out-erkrankungen/. Accessed 21 Oct 2019

https://bronnieware.com/blog/regrets-of-the-dying/. Accessed 21 Oct 2019

https://ggsc.berkeley.edu/what_we_do/event/the_science_of_happiness. Accessed 21 Oct 2019

https://storyofstuff.org/movies/. Accessed 21st Oct 2019

https://bmilab.com/topics/circular-economy

https://ec.europa.eu/eurostat/cache/infographs/energy/bloc-4c.html

https://www.sinus-institut.de/sinus-loesungen/sinus-milieus-deutschland/

http://designwithintent.co.uk. Accessed 21 Oct 2019

www.kickstarter.com. Accessed 21 Oct 2019

https://www.ecocrowd.de. Accessed 21 Oct 2019

https://www.leetchi.com. Accessed 21 Oct 2019

https://www.betterplace.org/de. Accessed 21 Oct 2019

Connecting Global Sustainability with Circular Product Design

3

Harald Desing

1 Introduction

Creating a sustainable society requires a fundamental transformation, which will affect all sectors and activities. Decarbonizing the economy and closing material cycles requires rethinking production and consumption in a systemic way, in order that individual actions actually contribute to a more sustainable future (Desing et al. 2020a, b, c). Products and service design has a pivotal role in creating a sustainable society, as it greatly influences the environmental and social impacts caused throughout multiple life cycles (Desing, 2021). In order to align design with sustainability, criteria for a sustainable society need to be defined.

Planet Earth is a finite entity, exchanging almost exclusively energy with space: incoming solar irradiation provides a large and steady low-entropy energy flux, which is balanced by outgoing long wavelength and high entropy radiation back to space (Szargut, 2003). Besides this "renewable" flux of energy, Earth is a closed system with a limited amount of resources. The majority of the incoming solar energy is utilized in Earth system processes, such as the water cycle, wind or the biosphere. Society is a part of this Earth system and is supported by it. However, humanity has become a driving force in the Earth system, where we massively intervene—mostly unintentionally—in Earth system processes through our technological abilities. For example, CO_2 emissions are not the desired output, but rather a side effect of providing energy from fossil resources. Yet, fossil energy use has increased atmospheric CO_2 concentration from 280 ppm before the industrial revolution to over 415 ppm today (Dlugokencky and Tans, 2022). We can design the system differently, in order to get what we need (e.g. energy) while

H. Desing (✉)
Empa, Technology and Society Laboratory, Lerchenfeldstrasse 5, 9014 St.Gallen, Switzerland
e-mail: Harald.desing@empa.ch

© The Author(s), under exclusive license to Springer Nature Singapore Pte Ltd. 2024
G. B. Melles and C. Wölfel (eds.), *Design for a Sustainable Circular Economy*, Design Science and Innovation, https://doi.org/10.1007/978-981-99-7532-7_3

minimizing the negative effects on the Earth system (e.g. CO_2, land use, nutrient fluxes). However, no technical intervention is free of environmental impacts: all need materials and energy, which are missing elsewhere in the Earth system (e.g. harvesting renewable energy diverts energy fluxes from formerly powered Earth system processes) and their mobilization is causing environmental impacts (e.g. from mining or land transformation). Consequently, scale matters! In a sustainable society, any physical activity is limited by Earth system boundaries—the aim and purpose of sustainable design is therefore to maximize services to society within this biophysical frame for human action. As our future prosperity is dependent on ecosystem services, it is in our own interest to maintain or even improve the integrity and functioning of the Earth system.

Primary extraction and final disposal of materials are responsible for major shares of environmental impacts (Desing et al. 2020a; IRP, 2019) and resource depletion (Henckens et al. 2014). Circulating materials, components and products in society can reduce impacts and preserve resources by increasing the utility of materials for society. This can be achieved by various circular strategies: extending the lifetime of entire products or parts thereof (e.g. reuse, remanufacturing, repair and repurpose), recycling the contained materials for the same or similar function by maintaining their quality and cascading materials to lower quality applications (e.g. crushed concrete as filling material). It is generally assumed that there is a "waste hierarchy" (European Commission, 2015), preferring, e.g. lifetime extension over recycling. However, each circular strategy requires energy and materials as well as causes environmental impacts depending on the system's design and performance. Consequently, there is no universal waste hierarchy, but the circular strategy with the best environmental performance has to be chosen in each individual case (Haupt and Hellweg, 2019; Hummen and Desing, 2021; Lama et al. 2022). This makes matters more complicated for designers because simple rules (such as "design durable products") cannot be applied. In contrast, it requires estimating or evaluating the environmental performance of circular strategies during the design process in order to make informed decisions and consistently minimize impacts (Desing et al. 2021a).

This chapter will outline how Earth system boundaries can be considered in sustainable circular design of products and services, connecting the global sustainability conditions with individual products and services and presenting a tool for estimating the environmental performance during the design process. This is based on a two-step procedure: first, environmental boundary conditions are translated into ecological resource budgets (Desing et al., 2020a, 2020b); and second, the consumption of limited resources is minimized within a product and service system using the resource pressure method (Desing et al., 2021a). The method can be used for both a relative comparison of circular design options as well as determining the contribution to absolute sustainability (Desing, et al., 2021b). Furthermore, a tool is introduced to help designers find the ecologically optimal replacement time and strategy for their product (Hummen and Desing, 2021).

2 Translating Earth System Boundaries into Resource Budgets

The planetary boundaries framework (Rockström et al., 2009; Steffen et al., 2015) proposes nine boundary categories for essential and critical Earth systems: climate, biodiversity, ozone depletion, ocean acidification, biogeochemical flows, land, freshwater use, aerosol loading and novel entities. Except for aerosol loading, global boundary values have been proposed (Persson et al., 2022). If none of these boundaries is violated, the Earth system is expected to remain in a Holocene-like state, similar to the very stable period of the last 10,000 years during which human civilizations emerged. A sustainable society will need to operate within these planetary limits, i.e. no boundary shall be violated.

Today, five out of the nine boundary categories are exceeded (Persson et al., 2022), where climate and biodiversity are the most critical (Desing et al., 2020a). Annual CO_2 emissions exceed the natural long-term sink of weathering and sedimentation by a factor of 50, while returning to a CO_2 concentration in the atmosphere of 350 ppm will require negative emissions at a massive scale (Desing et al., 2022). Reducing the pressure on the climate (i.e. decarbonization) has a synergetic effect on reducing pressure on biodiversity. This is, however, only if fossil energy is replaced primarily by solar energy conversion on the already built environment and not inducing further land transformation (Desing et al. 2019). Furthermore, increasing material circularity reduces the environmental impacts caused by primary material extraction and final disposal. In addition to transforming the provisioning system, the pressure on Earth system boundaries can be decreased by reducing consumption (Reike et al. (2017) term this the circular strategies "reduce" and "rethink").

Considering environmental boundaries in design is not straightforward, because design only indirectly influences impacts. However, designers have direct control over resource consumption, as the design determines the quality and quantity of resources required in the product and service. In order to connect environmental impacts with resource consumption, global boundaries have to be translated into ecological resource budgets.

There are two approaches to calculating resource budgets: (i) Ecological Resource Availability (ERA) (Desing et al., 2020a) calculates the maximum sustainable annual production of all resources so that globally none of the considered boundaries is violated, and (ii) Ecological Resource Potential (ERP) (Desing et al., 2020b) estimates the theoretical potential of annual production not exceeding any global boundary for one resource in isolation. ERA allows us to make an absolute assessment of the sustainability of resource consumption. However, it requires normative allocation of global boundaries to individual resources, as the global environmental space has to be shared by all societal activities. ERP, in contrast, asks the question: "what is the maximum production of a resource if no other activity would take place?" As such, it is an upper limit (the ecological potential) in the hypothetical case when all other resources would be substituted by it. It allows us to compare different resources relative to each other based on the most limiting boundary category for each resource. For example, Al production is limited by CO_2 emissions, while Cu is limited by biodiversity. Cu has much lower climate impacts than Al, however, it causes much higher biodiversity loss. ERP is an objective comparison considering multiple impact categories without the need for subjective and normative weighting.

Both ERP and ERA provide results for a specific provisioning system, i.e. impacts generated in a particular supply chain for producing the material. Such a provisioning system can be modelled as the current (unsustainable) economy (such as represented by life cycle inventory databases, e.g. ecoinvent (Wernet et al., 2016) or environmentally extended input–output tables, e.g. exiobase (Stadler et al., 2018)) or any future scenario during the transformation (Mendoza Beltran et al., 2018).

For a relative comparison of different product designs and circular strategies, the ERP method provides a solid reference for considering all impact categories relevant to Earth system boundaries (Lama et al., 2022). Resources are preferred, if they have high ERP and thus low impacts on all boundaries. For an absolute sustainability analysis, ERA budgets need to be calculated for a specified allocation narrative and provisioning system scenario (Desing et al., 2021b; Ryberg et al. 2020).

The calculation of ERA requires multiple steps (Desing et al. 2020a):

- Define Earth system boundaries (ESB), control variables and the maximum probability of violation for any boundary (P_V).
- Define a single resource or resource segment to investigate; segments are required to specify the relative share of production (SoP) of resources contained in the segment.
- Allocate fractions of the global boundaries to the resource or resource segment (share of safe operating space, SoSOS).
- Calculate the impacts on ESB for one unit of resource production and end-of-life treatment using a life cycle assessment approach. It is important to include the impacts from final disposal (e.g. dispersion in the environment, landfilling, sewage or incineration), as they can be substantial (e.g. CO_2 emissions from incinerating plastics are about the same as for primary production) and are associated with the linear use of materials.
- Determine ERA: Upscale production of resources within a segment until impacts exceed the most limiting allocated boundary with the chosen probability of violation.

The calculation of ERP builds on the structure established for ERA (Desing et al. 2020b):

- Define Earth system boundaries (ESB), control variables, and the maximum probability of violation for any boundary (P_V).
- Calculate the impacts on ESB for one unit of resource production and end-of-life treatment using a life cycle assessment approach.

– Determine ERP: upscale production of this resource until impacts exceed the most limiting global boundary with the chosen probability of violation.

Table 1 provides an overview of ERP results obtained with ecoinvent v3.6, which represent the current and—depending on the datasets—past provisioning system. The larger the ERP, the lower the impact on ESB. Note that most materials are limited by CO_2, however, some are limited by N emissions and Cu is limited by biodiversity impacts. When decarbonizing the provisioning system, it can be expected that other boundaries will become limiting, e.g. land use related boundaries when shifting towards a bio-economy.

3 Resource Pressure

Products and services require resources (energy and materials) to provide functions to society. Design determines the quantity and quality of resources necessary throughout the whole life cycle. As most impacts on Earth system boundaries are generated by the production and disposal of resources, it is pivotal to reduce their consumption. The resource pressure method (Desing et al., 2021a) provides the designer (and other decision makers) with an indicator and design guidelines to consider ecological resource budgets during the design and evaluation of circular strategies. The method tracks the mass flow of a material through the product/ service system.

In addition to the material contained in the product ($m_{product}$), a surplus is necessary to make up for manufacturing losses ($\gamma_{manufacturing}$). The mass flow to provide the continuous function over time is the mass necessary to produce the product divided by the lifetime t_L:

$$\dot{m}_{product} = \frac{m_{product}}{t_L}\left(1 + \gamma_{manufacturing}\right).$$

This mass flow can be satisfied by three different sources: primary material, recycled material from the same product system and cascaded material from a different product system. The distinction between recycling and cascading is important, as the former requires maintaining the quality of the material to be able to use it for the same function again. Cascading, in contrast, allows the utilization of material further at lower quality. At the end of life, the material can again be split into three categories: final waste that is "unrecoverable", i.e. recovery is only possible at energy demand and efforts higher than for primary material extraction; cascading to a lower grade purpose and recycling for the material, which can be recovered at the same quality necessary to provide the same function again. The recycling flow is characterized by the parameter recyclability $\eta_r = \frac{\dot{m}_r}{\dot{m}_{product}}$, greatly influenced by design decisions but also recycling system performance (Reuter and van Schaik, 2015). A material can be considered recyclable if it can be applied for its original function. To enable recyclability, materials need to remain homogenous and easy to separate during recycling from other materials. For example, a

Table 1 ERP for various materials calculated with ecoinvent 3.6

Resource	ERP / kg/a	Limiting boundary
Concrete, 20MPa	7.18E + 12	CO_2
Concrete, 25MPa	8.46E + 12	CO_2
Concrete, 30-32MPa	6.23E + 12	CO_2
Concrete, 35MPa	6.67E + 12	CO_2
Concrete, 50MPa	4.90E + 12	CO_2
Clay brick	2.77E + 12	CO_2
Light clay brick	5.77E + 12	CO_2
Shale brick	3.81E + 12	CO_2
Sanitary ceramics	4.95E + 11	CO_2
Ceramic tile	1.13E + 12	CO_2
Flat glass, uncoated	8.65E + 11	CO_2
Packaging glass, brown	7.05E + 11	CO_2
Packaging glass, green	7.05E + 11	CO_2
Packaging glass, white	7.06E + 11	CO_2
Glass wool mat	2.94E + 11	CO_2
Aluminium, primary	4.14E + 10	CO_2
Copper primary	5.60E + 10	Biodiversity
Steel, unalloyed	4.37E + 11	CO_2
Cast iron	5.72E + 11	CO_2
Zinc	2.34E + 11	N emissions
Lead	1.53E + 11	N emissions
Tin	9.01E + 10	CO_2
Nickel, 99.5%	6.85E + 10	CO_2
Gold	1.89E + 07	CO_2
Silver	1.82E + 09	CO_2
Platinum	1.33E + 07	CO_2
Titanium, primary	3.06E + 10	CO_2
Chromium	3.51E + 10	CO_2
Steel, chromium steel 18/8, hot rolled	1.96E + 11	CO_2
Polystyrene, general purpose	2.06E + 11	CO_2
Polyethylene terephthalate, granulate, bottle grade	2.25E + 11	CO_2
Fibre, polyester	1.70E + 11	CO_2
Polyvinylchloride, bulk polymerised	2.37E + 11	CO_2
Polyethylene, high density, granulate	2.63E + 11	CO_2
Polyethylene, low density, granulate	2.52E + 11	CO_2
Polypropylene, granulate	2.68E + 11	CO_2

(continued)

Table 1 (continued)

Resource	ERP / kg/a	Limiting boundary
Polypropylene fibre	2.05E + 11	CO_2
Polycarbonate	1.13E + 11	CO_2
Polyurethane	1.53E + 11	CO_2
Nylon 6	1.23E + 11	CO_2
Nylon 6–6	1.07E + 11	CO_2
Polyacryl	1.74E + 11	CO_2
Fibre, viscose	1.90E + 11	CO_2
Fibre, cotton	2.45E + 11	N emissions
Fibre, cotton, organic	4.96E + 11	N emissions
Sheep fleece in the grease	1.04E + 11	N emissions
Reeled raw silk hank	9.17E + 09	CO_2
Fibre, jute	1.09E + 12	CO_2
Fibre, flax	7.75E + 11	CO_2
Acrylonitrile–butadiene–styrene copolymer	2.23E + 11	CO_2

part made from low-alloyed Al can be remoulded to the same part with minimal losses, if it is separated from the product, cleaned and directly treated in a dedicated process. However, in most recycling systems, Al parts are not separated from products and are shredded together with the whole product. Due to incomplete sorting and mechanical entanglement, the Al fraction is contaminated with other metals, allowing for high-alloyed cast applications only. In the latter case, recyclability is zero, while in the former case, it is close to one. Materials that cannot be recycled, may qualify for cascading (as in the example of Al cast alloys above). This material flow, characterized by cascadability $\eta_c = \frac{\dot{m}_c}{\dot{m}_{product}}$, is not lost for society, however, has lost quality. Cascading can only be considered, if there is a (large enough) market for the material; otherwise, material would go to final waste. Cascaded material as an input is considered a "secondary material" input and can be chosen by the designer if specifications allow and the secondary material is available on the market. The primary material input is specified through the primary material content $\alpha = \frac{\dot{m}_p}{\dot{m}_{product}}$ or the modified primary material input, which specifies the primary material fraction of inputs excluding recycling: $\alpha\prime = \frac{\dot{m}_p}{\dot{m}_p + \dot{m}_s}$. The two values are related with $\alpha = \alpha\prime(1 - \eta_r)$.

The resource pressure τ is defined as the fraction of primary resource consumption induced by the product (or service) system in relation to the respective ecological resource budget. Resource pressure is thereby exerted in two ways: (i) by consuming primary resources directly and (ii) by generating final losses, which induce primary material production elsewhere in the socio-economic metabolism.

The resource pressure is defined as the weighted average between (i) and (ii):

$$\tau = \frac{1}{2}\frac{\dot{m}_p + \dot{m}_{loss}}{ERB} = \frac{1}{2}\frac{m_{product}}{ERB}\frac{1}{t_L}(1 + \gamma_{manufacturing})(1 + \alpha - \eta_r - \eta_c).$$

The resource pressure for multiple resources can simply be added up to arrive at a value for the entire product. It is a dimensionless number, which has no direct interpretation when using ERP but is the fraction of the maximum sustainable global resource production ERA necessary to provide the intended product or service. In the latter case, the sum of all resource pressures for a specific resource would need to be $\Sigma\tau < 1$ in order to be absolutely sustainable (Desing et al., 2021b).

For energy use in the product system (i.e. for manufacturing, use or recovery), there is only cascading (e.g. use of waste heat) but no recycling possible (every use of energy reduces the exergy content and increases entropy, second law of thermodynamics).

From the resource pressure indicator, several design guidelines can be derived, which help considering the environmental dimension of circularity already during design conception (Desing et al., 2021a):

- Choose materials with large ERB, i.e. with low environmental impacts.
- Reduce mass in the product.
- Minimize manufacturing losses.
- Increase lifetime.
- Reduce primary material input.
- Increase recyclability.
- Increase cascadability.

While these general guidelines are indicative for a single resource, they are not necessarily applicable to the entire product system. Consider a product with a significant impact during the use phase, such as a gas boiler. Here it may be beneficial to replace the product prematurely with a more efficient alternative. In the case of gas boilers, the environmentally optimal lifetime is zero when considering a replacement with a heat pump, i.e. it is environmentally beneficial to make a gas boiler waste and recover its materials (Hummen and Desing, 2021). For this reason, the environmental lifetime optimizer has been developed (Hummen and Desing, 2021), providing guidance to product designers regarding the optimal time and strategy to replace products with significant impacts in the use phase.

Similarly, also recycling and cascading need energy and cause impacts, thus there are environmental optima for recyclability and cascadability (Schäfer, 2021; Schmidt, 2021). Up to today, there are no tools available to guide the designer in the question of which level of recycling and cascading to target, which is a potential area for future research.

Various case studies show a good agreement between the resource pressure results and life cycle assessment (LCA) scores across a wide range of impact categories (Desing et al., 2021a; Desing et al., 2021b; Lama et al., 2022). At the same time, the resource pressure method is much easier to apply and does not require expert knowledge on environmental assessments, as will be showcased in the next section.

4 Illustrative Case Study

To illustrate the approach described before, let's consider two different beverage can designs and multiple circular strategies. One can is made from Al, the second from steel. ERP of steel is one order of magnitude larger than for Al. The aluminium can weights 13 g, while the steel weights 22 g. Manufacturing losses are considered fully recyclable and excluded for simplicity. Cascadability as well as secondary material input are set to zero. The lifetime for each can is set for simplicity as 1a. Note that the lifetime depends on the shelf life and can be considered the same for Al and steel, i.e. any other lifetime leads to the same relative results. Recycling is costly in terms of energy: Al recycling requires 0.71 kWh/kg (Boin and Bertram, 2005), while steel recycling requires 0.4 kWh/kg (Haupt et al. 2017). Similar to materials, also energy demand exerts resource pressure. The ERP of energy resources are listed in Table 2.

The baseline scenario considers a collection rate of 75% and steady state recyclability of 80% for Al (Lovik and Müller, 2014) and 90% for steel, respectively (United Nations Environment Programme, 2011). The resource pressure is

Table 2 ERP for electricity provided with different technologies based on ecoinvent 3.9; biophysical limits are taken from Desing et al. (2019)

Electricity provision	ERP kWh/a	Limiting boundary
Rooftop PV	2.81E + 13	CO_2
Wind onshore	1.13E + 12	Biophysical
Wind offshore	9.84E + 11	Biophysical
Wood	1.12E + 12	Biophysical
Concentrated solar power in deserts	2.02E + 13	CO_2
Deep geothermal	2.55E + 12	Biophysical
Coal	7.57E + 11	CO_2
Natural gas	1.42E + 12	CO_2
Oil	8.80E + 11	CO_2
Nuclear, boiling water reactor (CH)	7.03E + 12	CO_2
Global electricity average	1.26E + 12	CO_2
Global electricity average for aluminium industry	1.13E + 12	CO_2

calculated for Al material contained in the can:

$$\tau_{Al,can} = \frac{1}{2}\frac{m_{product}}{ERB}\frac{1}{t_L}(1+\gamma_{manufacturing})(1+\alpha\prime(1-\eta_r)-\eta_r-\eta_c) = \frac{1}{2}\frac{0.013kg}{4.14\times10^{10}\frac{kg}{a}}\frac{1}{1a}(1+0)(1+1(1-0.6)-0.6-0) = 1.26\times10^{-13}.$$

The resource pressure of energy necessary for recycling the recycling mass flow (primary content in electricity is one) is calculated as follows:

$$\tau_{Al,E,r} = \frac{1}{2}\frac{E_r\dot{m}_r}{ERB}(1+\alpha) = \frac{E_r}{ERB}\frac{m_{product}\eta_r}{t_L} = \frac{0.71\frac{kWh}{kg}}{1.13\times10^{12}\frac{kWh}{a}}\frac{0.013kg0.6}{1a} = 4.9\times10^{-15}.$$

The total resource pressure is in the baseline scenario dominated by the aluminium in the can and amounts to $\tau_{Al} = 1.3\times10^{-13}$. For the steel can, recycling energy exerts about the same resource pressure ($\tau_{st,E,r} = 4.7\times10^{-15}$), but the resource pressure of the steel can is one order of magnitude lower than Al can ($\tau_{st,can} = 1.6\times10^{-14}$). Recycling contributes about 1/5 of total of $\tau_{steel} = 2.1\times10^{-14}$ (see Fig. 1).

In the R + scenario, the recyclability is increased through an improved collection to 95%. This reduces the resource pressure for both Al and steel by about 40% each, while it doubles the relative contribution of recycling energy.

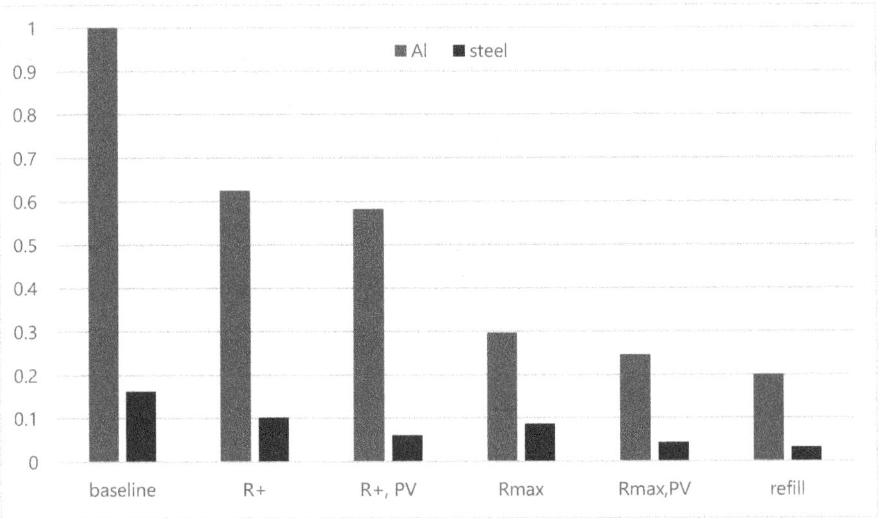

Fig. 1 Case study results for resource pressure relative to Al-baseline scenario for beverage cans made from Al and steel

When considering solar rooftop PV as the energy resource for recycling in the R + scenario, the resource pressure of recycling energy is reduced by one order of magnitude. This decreases the resource pressure for Al by only 7%, while again by 40% for steel. As the steel can contribute much less to the overall resource pressure, the effect of powering recycling with PV is much more pronounced, even though the resource pressure for both Al and steel recycling improves by the same total value.

Setting the recyclability to a maximum of 90% (Rmax), the resource pressure can be halved for Al in comparison to R + , however, this only improves by 15% for steel. This is due to the fact that the resource pressure of Al is dominated by material. Consequently, improving the recyclability of the material reduces the resource pressure much more than the effect of the increased energy demand. For steel, recycling energy is more important, thus increasing recyclability also increases the resource pressure for recycling energy and compensates most of the gains in the material's resource pressure. Powering recycling with PV (Rmax,PV) improves the situation for steel, yet Al still benefits more from the increased recyclability.

Finally, testing the effect of increased lifetime, the refill scenario considers reusing cans 10 times before recycling them in the same way as in the baseline scenario. The refillable cans need to be more sturdy, thus simply assuming double mass contained in each can. This reduces resource pressure fivefold compared to baseline and is by far the best scenario. Note that for simplicity, energy for collection and washing is disregarded, however, it can be considered minor compared with the energy for recycling (i.e. melting, refining and fabricating metal containers).

In conclusion, steel cans perform much better than Al cans despite the fact that it is heavier because steel has an ERP larger by one order of magnitude. Additionally, steel is better recyclable and requires less energy for recycling. However, the ERP of Al may increase significantly, if primary Al is produced with fossil-free energy in a decarbonized provisioning system. Recycling can reduce the resource pressure substantially, yet, parts of the potential improvement are reversed by increased energy demand for recycling. Refilling has an even greater benefit than recycling: preserving the integrity of the can saves most of the environmental impacts associated with recycling and repeated production.

This case study shall illustrate the utility of the resource pressure method. Many aspects, variants and scenarios can be investigated more in detail, e.g. including energy for transport or washing and sorting, and cascading the use of materials for and from cans. All these open questions can be addressed by more detailed analysis following the same principles as showcased here.

5 Outlook

The resource pressure approach can be applied in the design and development practice at companies without the need for special expertise. It may be combined with ex-post LCA of the finished product to establish checks of the attained environmental performance and provide feedbacks to the design teams. Furthermore, it can be used as a screening tool for promising circular business models, new markets and technological avenues. As mentioned earlier, absolute sustainability assessments are possible when using ERA budgets. These depend on the allocation scenarios chosen for the calculation of ERA. The resource reduction index (Desing et al., 2021b) provides a framework to assess the contribution of circular strategies on reaching global sustainability. It measures the achievement of reduction targets for resource consumption within the product system, where the targets are set based on global consideration regarding sustainable resource use in a sector or industry. To further operationalize this approach, allocation principles and reduction targets have to be defined for reaching environmental sustainability and satisfying beyond universal basic needs (i.e. for reaching a doughnut economy (Raworth, 2012)). As design determines to a large extent recycling and cascading performance (Reuter et al. 2019), lifetime (Den Hollander, 2018) and use behaviour (Srivastava and Shu, 2013), there is a need to develop practical tools to guide design decisions towards optimal circularity (Schmidt, 2021).

References

Boin UMJ, Bertram M (2005) Melting standardized aluminum scrap: a mass balance model for europe. JOM 57(8):26–33. https://doi.org/10.1007/s11837-005-0164-4

Desing H, Widmer R, Beloin-Saint-Pierre D, Hischier R, Wäger P (2019) Powering a sustainable and circular economy—an engineering approach to estimating renewable energy potentials within earth system boundaries. Energies 12(24):1–18. https://doi.org/10.3390/en12244723

Desing H, Braun G, Hischier R (2020a) Ecological resource availability: a method to estimate resource budgets for a sustainable economy. Glob Sustain 3:1–11. https://doi.org/10.1017/sus.2020.26

Desing H, Braun G, Hischier R (2020b) Ecological resource potential. MethodsX 7. https://doi.org/10.1016/j.mex.2020.101151

Desing H, Brunner D, Takacs F, Nahrath S, Frankenberger K, Hischier R (2020c) A circular economy within the planetary boundaries: towards a resource-based, systemic approach. Resour Conserv Recycl 155 https://doi.org/10.1016/j.resconrec.2019.104673

Desing H, Braun G, Hischier R (2021a) Resource pressure–a circular design method. Resour Conserv Recycl 164 https://doi.org/10.1016/j.resconrec.2020.105179

Desing H, Braun G, Hischier R (2021b) The resource reduction index–evaluating product design's contribution to a sustainable circular economy. Paper presented at the European Roundtable on Sustainable Consumption and Production 2021, Graz, Austria

Desing H, Gerber A, Hischier R, Wäger P, Widmer R (2022) The 3-machines energy transition model: exploring the energy frontiers for restoring a habitable climate. Earth's Future 10(10):1–15. https://doi.org/10.1029/2022ef002875

Desing H (2021) Product and service design for a sustainable circular economy. (Ph.D.), ETH Zürich, Zürich (27225)

Den Hollander M (2018) Design for managing obsolescence-a design methodology for preserving product integrity in a circular economy. TU Delft
Dlugokencky E, Tans P (2022) NOAA trends in atmospheric carbon dioxide. Retrieved from www.esrl.noaa.gov/gmd/ccgg/trends/
European Commission (2015) Closing the loop-an EU action plan for the ciruclar economy. Retrieved from
Haupt M, Vadenbo C, Zeltner C, Hellweg S (2017) Influence of input-scrap quality on the environmental impact of secondary steel production. J Ind Ecol 21(2):391–401. https://doi.org/10.1111/jiec.12439
Haupt M, Hellweg S (2019) Measuring the environmental sustainability of a circular economy. Environ Sustain Indic 1–2. https://doi.org/10.1016/j.indic.2019.100005
Henckens M, Driessen P, Worrell E (2014) Metal scarcity and sustainability, analyzing the necessity to reduce the extraction of scarce metals. Resour Conserv Recycl 93:1–8. https://doi.org/10.1016/j.resconrec.2014.09.012
Hummen T, Desing H (2021) When to replace products with which (circular) strategy? An optimization approach and lifespan indicator. Resour Conserv Recycl 174. https://doi.org/10.1016/j.resconrec.2021.105704
IRP (2019) Global resources outlook. Retrieved from Nairobi, Kenya. https://www.resourcepanel.org/reports/global-resources-outlook
Lama V, Righi S, Quandt BM, Hischier R, Desing H (2022) Resource pressure of carpets: guiding their circular design. Sustainability 14(5):2530. https://doi.org/10.3390/su14052530
Lovik AN, Müller DB (2014) A material flow model for impurity accumulation in beverage can recycling systems. Light Metals 907–911
Mendoza Beltran A, Cox B, Mutel C, Vuuren DP, Font Vivanco D, Deetman S, Tukker A (2018) When the background matters: using scenarios from integrated assessment models in prospective life cycle assessment. J Ind Ecol 24(1):64–79. https://doi.org/10.1111/jiec.12825
Persson L, Carney Almroth BM, Collins CD, Cornell S, de Wit CA, Diamond ML, Hauschild MZ (2022) Outside the safe operating space of the planetary boundary for novel entities. Environ Sci Technol 56(3):1510–1521. https://doi.org/10.1021/acs.est.1c04158
Raworth K (2012) A safe and just space for humanity: can we live within the doughnut. Oxfam Policy Pract: Clim Chang Resil 8(1):1–26. https://www.ingentaconnect.com/content/oxpp/oppccr/2012/00000008/00000001/art00001
Reike D, Vermeulen WJV, Witjes S (2017) The circular economy: new or refurbished as CE 3.0?—exploring controversies in the conceptualization of the circular economy through a focus on history and resource value retention options. Resour Conserv Recycl 135:246–264. https://doi.org/10.1016/j.resconrec.2017.08.027
Reuter M, van Schaik A (2015) Product-centric simulation-based design for recycling: case of LED lamp recycling. J Sustain Metall 1:4–18. https://doi.org/10.1007/s40831-014-0006-0
Reuter M, van Schaik A, Gutzmer J, Bartie N, Abadías-Llamas A (2019) Challenges of the circular economy: a material, metallurgical, and product design perspective. Annu Rev Mater Res 49(1):253–274. https://doi.org/10.1146/annurev-matsci-070218-010057
Rockström J, Steffen W, Noone K, Persson A, Chapin S, Lambin E, Foley J (2009) Planetary boundaries: exploring the safe operating space for humanity. Ecol Soc 14(2):1–33
Ryberg MW, Andersen MM, Owsianiak M, Hauschild MZ (2020) Downscaling the planetary boundaries in absolute environmental sustainability assessments–a review. J Clean Prod 276. https://doi.org/10.1016/j.jclepro.2020.123287
Schäfer P (2021) Recycling-ein Mittel zu welchem Zweck? Modellbasierte Ermittlung der energetischen Aufwände des Metallrecyclings für einen empirischen Vergleich mit der Primärgewinnung. Springer Spektrum, Germany
Schmidt M (2021) The resource-energy nexus as a key factor for circular economy. Chem Ing Tec 93(11):1707–1716. https://doi.org/10.1002/cite.202100111
Srivastava J, Shu LH (2013) Encouraging resource-conscious behavior through product design: the principle of discretization. J Mech Des 135(6). https://doi.org/10.1115/1.4024225

Stadler K, Wood R, Bulavskaya T, Södersten C-J, Simas M, Schmidt S, Tukker A (2018) EXIOBASE 3: developing a time series of detailed environmentally extended multi-regional input-output tables. J Ind Ecol 22(3):502–515. https://doi.org/10.1111/jiec.12715
Steffen W, Richardson K, Rockstrom J, Cornell SE, Fetzer I, Bennett EM, Sorlin S (2015) Sustainability. Planetary boundaries: guiding human development on a changing planet. Sci 347(6223):1259855. https://doi.org/10.1126/science.1259855
Szargut JT (2003) Anthropogenic and natural exergy losses (exergy balance of the earth's surface and atmosphere). Energy 28(11):1047–1054. https://doi.org/10.1016/s0360-5442(03)00089-6
United Nations Environment Programme (2011) Recycling rates of metals-a status report. A report of the working group on the gloal metal flows to the international resource panel. Retrieved from
Wernet G, Bauer C, Steubing B, Reinhard J, Moreno-Ruiz E, Weidema B (2016) The ecoinvent database version 3 (part I): overview and methodology. Int J Life Cycle Assess 21(9):1218–1230. https://doi.org/10.1007/s11367-016-1087-8

Mapping Circular Economy and the Role of Design in Portugal

4

Nicola Morelli, Nina Costa, and Luca Simeone

After a long period of maturation, the European Commission approved an action plan for Circular Economy in 2015 (EC, 2015). The plan was received and approved by the EU countries, who followed up with national plans. Once the policies have been set, it is also important to recognize the concrete changes they are triggering in the socio-economic context, and, from the perspective of a designer, to understand the possible role of design in such change. This suggests two research questions that can be analyzed in relation to CE initiatives in Portugal. The Portuguese action plan sets the actions for CE at the macro, meso, and micro levels, thus including actions that concern broad strategic and cultural changes (at the macro level), and identify key sectors (at the meso level) and local actions (at the micro level). The authors of this paper analyzed the circular economy initiatives in Portugal, that have been recorded in a database issued by the Circular Economy Club. The shift from a linear to a circular economy culture and practice can be read through a classification of different initiatives (Alaerts et al., 2019), which helps in recognizing and qualifying the dimension of change. The analytical framework proposed in this paper also considers a Multi-Level Perspective (Geels and Schot, 2007), which allows the inclusion of some initiatives that could not be framed in Alaerts' classification. The change generated by the new policy extends designers' role to new functions: while the potential for circularity in product design has been largely assimilated into designers' activity, the new rules and the panorama of the ongoing initiatives suggest a possible critical role

N. Morelli (✉) · L. Simeone
Aalborg University, Copenhagen, Denmark
e-mail: nmor@create.aau.dk

N. Costa
Research Institute of Design, Media and Culture (ID+), University of Aveiro, Aveiro, Portugal

© The Author(s), under exclusive license to Springer Nature Singapore Pte Ltd. 2024

G. B. Melles and C. Wölfel (eds.), *Design for a Sustainable Circular Economy*,
Design Science and Innovation, https://doi.org/10.1007/978-981-99-7532-7_4

for designers to infrastructuring change. The inclusion of design in policy actions and their role in translating policy actions into concrete initiatives at the local level is something relatively new, that opens new perspectives and new challenges.

1 Back to the Roots

At the beginning of the last decade, the Ellen McArthur Foundation (MacArthur, 2013) defined the Circular Economy (CE from now on) as a complex systemic approach to sustainable production and consumption; although the CE has become a "hot" topic for discussion in the last decade, it represents the maturation of different streams of research on environmental strategies and sustainability in the last 30 years. In 1992, Robert Frosh (Frosch, 1992), for example, proposed that a strategic approach toward a sustainable transformation of industrial ecosystems could be based on an analogy between natural cycles and industrial cycles.

In nature, the concept of *waste* does not exist as every material that is discarded by an organism, or a natural process becomes a nutrient for other organisms or natural processes. By proposing the concept of *Industrial Ecology*, Frosh opened a new perspective on industrial systems. The concept was taken up by Ayres and Simonis (Ayres and Simonis, 1994) to highlight the circular metabolic processes that industrial processes could adopt, imitating nature, to reduce the amount of industrial waste. The concept of metabolism is used to indicate the capability of natural ecosystems to *digest* waste from natural processes, transforming them into resources for new processes. Frosch and Gallopoulos (1989) used this concept for industrial systems and proposed that waste from one process could serve as raw material from another, thus reducing the overall industrial impact on the environment.

The concepts of industrial ecology and *industrial metabolism* were a good theoretical foundation in the debate on *Extended Producer Responsibility* (EPR), which extends the responsibility of companies over their own products beyond the point of sale (Gertzakis et al., 2002). Extended Producer Responsibility was introduced in Germany in 1991 as a waste management measure concerning packaging (Cahill et al., 2011); soon other countries followed the initiative, and other sectors, from automotive to photocopiers, seized the need for a deep rethinking not only in the business strategies, but also in the way products were designed (John Gertsakis et al., 1998).

The forced need to reuse the material discarded from the use of product opened a perspective over the whole life cycle and the concept of industrial metabolism highlighted the systemic nature of the initial idea proposed by Frosh. Material cycles could not be limited to a specific industrial sector, in which the material had been produced, but had rather to be extended to the various sectors that could be included in the production and consumption cycle. The life cycle perspective suggested a shift from a linear approach (*from cradle to grave*) to a circular perspective (*from cradle to cradle*) (Braungart and McDonough, 2008).

2 From Industrial Systems to Production and Consumption Loops

This short summary of the evolution of different streams of thought around the roots of circular economy reveals that the theoretical ground for rethinking industrial production systems was quite solid at the end of the last century. However, the implementation of the theoretical formulation shows a level of complexity that is still being debated in several aspects, such as the possibility of decoupling economy and growth, and consequently the capability of capitalism to address CE within its own paradigmatic structure (Melles, 2021), or the implementation of CE in urban contexts (Campbell-Johnston et al., 2019). The analysis of products' life reveals that the conditions for circularity were different from product to product: some products could be easily reused or repaired, and others could be remanufactured (Giarini and Stahel, 1986; Stahel, 1994), thus creating the conditions for product life extension. Others (like cars and white appliances) could be replaced by new products, given the higher efficiency of the new models in respect to the old ones (Heiskanen, 1996; Stahel, 1994).

The Ellen McArthur Foundation's butterfly diagram (MacArthur, 2013) includes several strategies to manage the stock of resources, including natural resources and those resources that have already been embedded in the production and consumption system (Fig. 2). The strategies on the right side of the diagram focus on products' material components from the perspective of recirculating resources. In this scheme, we could read different levels of intervention, based on the quality of knowledge required to feed the cycle: the larger cycles imply an intervention on the *micro* level, i.e. on materials. Sometimes this intervention concerns the molecular manipulation of material, as for polymeric materials, or mechanical treatments, as for organic materials or simply the destruction of molecular links, as for energy recovery. This intervention usually reduces the value and quality of materials and requires advanced and specialized, technical knowledge, besides the specific knowledge needed for the activation of material flows and the adaptation of business models. Progressing toward the narrower cycle, strategies are referring to larger systemic change, which relies on business or socio-cultural knowledge. If we consider the technical knowledge embedded in products (from molecular to mechanical knowledge), the strategies related to the narrower cycles in Fig. 1 are those that recover most of the knowledge embedded in products, whereas the larger cycles refer to strategies that imply a higher use of resources (to transport, disassemble and recycle products) and the larger waste of technical knowledge (downcycling). Indeed material recycling or incineration implies that all the knowledge embedded in a product is destroyed and only the minimal characteristics of the material or the pure energy is recovered.

While the circular use of resources can be easily monitored in larger circles, the narrower circles require a strategic approach that cannot be exemplified through the logic of flows, including strategies to reduce energy use, increase usage intensity, and possibly consider different functions to address specific needs.

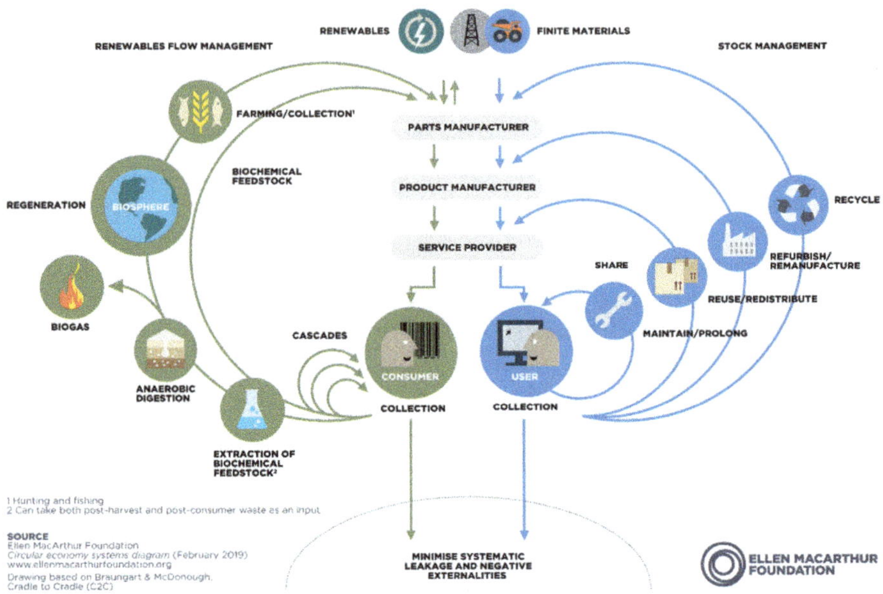

Fig. 1 The circular economy system diagram (*Source* McArthur foundation)

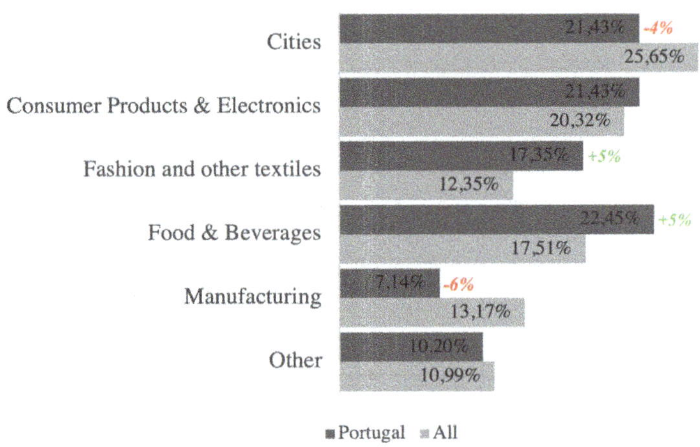

Fig. 2 CE initiatives by sector

Alaerts et al. (Alaerts et al., 2019) classify circularity strategies from linear to circular economy, taking into consideration the complexity of the initiatives, which require a progression from technological innovation to socio-technical change. By doing this, they consider strategies for smarter product use and manufacture, which include strategies for addressing users' needs through a range of different functions or to increase the usage intensity or product efficiency.

The different strategies proposed in this case are focused on product use and implicitly indicate a logical progression from linear processes, mainly connected with technological innovation initiatives, to socio-technical change (see Table 1). This progression excludes the purely linear processes—those in which resources are extracted from nature, used, and discarded in nature—but starts from the simple recovery of materials (level C), which corresponds to the larger circles of the butterfly diagram in Fig. 1, which focus on materials and energy recovery. The strategies increase their complexity when they imply services, or institutional infrastructure or socio-cultural changes (level B and A) that involve a larger part of the socio-technical system.

Some authors highlighted the systemic aspects related to the strategies for product life extension. In particular, Stahel (Stahel, 1997) proposed a *service (cycle) economy*, in which the closure of material loops was proposed in relation to the closure of *liability loops*, i.e. the creation of services to distribute the liabilities/responsibilities in a society working on closed material and energy loops. The

Table 1 Circularity strategies, adapted from (Alaerts et al., 2019)

Circular economy	Level A smarter product use and manufacture	Refuse	Make the product redundant by shifting function	Socio-technical change
		Rethink	Increasing usage intensity	
		Reduce	Increasing product efficiency by using fewer resources	
Linear economy	Level B extend the lifespan of products or parts	Re-use	Reusing a discarded product that is still in good condition	Technological innovation
		Repair	Repair or maintain defective product	
		Refurbish	Restore and old product bringing it up to date	
		Remanufacture	Re-use part of a discarded product in a new product with the same function	
		Repurpose	Use a discarded product or its parts for a new function	
	Level C useful application of materials	Recycle	Process materials to obtain new material	
		Recover	Energy recover	

service analyzed in Stahel's exploration included infrastructure for sharing the utilization of goods or recycling/remanufacturing goods as well as services that referred to systemic socio-cultural solutions, such as platforms (the term platform was not used by Stahel, as it was introduced a few years later) to sell results, utilization of goods, in a perspective of a *functional economy* (Giarini and Stahel, 1986) or *Performance Economy* (Stahel, 2010).

The principles of Extended Producer Responsibility and the idea of an extension of the concept of industrial metabolism to larger human systems (such as cities or regions) are translated into the concept of circularity and mapped onto various concrete aspects of existing socio-technical systems, from product-related strategies aimed at facilitating product disassembly and material recycling (Gertzakis et al., 2002) to product life extension, which often implies the revision of the principle of product ownership (Stahel, 1997). This debate was extended to a broader discussion between those who call for a total redesign of the whole socio-technical system, based on capitalist principles, and those who believe in the capability of the same system to address CE with a number of adjustments (Melles, 2021).

3 A Multilevel Perspective on Sustainability Transition

The perspective shift from products to their utilization is part of the debate on the nature of the change proposed as the basis for the circular economy: the initial perspective focused on products suggests strategies to extend product life (through repair or modular configuration) or the life of materials (through product disassembly and recycling). This perspective was mostly geared toward localized changes in product configurations and design. The idea of extending the producer's responsibility to the whole product lifecycle, though, implied that no substantial change was possible without the support of an infrastructural system.

In this sense, the systemic analysis of the change required by a perspective of circular use of products can be framed through multilevel sustainability transition theory proposed by Geels and Schot 2007): the perception of a risk connected to the extension of product responsibilities led companies to generate different product-related strategies, but such strategies were fated to stay in a niche, they could only emerge and be extended to their entire industrial sector when *regime conditions* (e.g. policy-driven initiatives, that activate specific services, infrastructures, legal configurations) could be created, which could facilitate the operation of maintenance, repair, and recirculation of products, components, and materials (Melles, 2021).

The emergence of such regime conditions, in turn, was only possible once a socio-cultural change forced a substantial shift in the focus of public debate, legislation, and institutional structures. The definition of such a shift implies cultural changes at different levels. Some authors, such as Stahel (Stahel, 1997), proposed a shift from the intrinsic value of a product to the value of its utilization. The implications of such shift are cultural, before being technical, as they concern not only single products, but social practices and lifestyles, ways of thinking and perceiving reality, i.e. what Geels and Schot define as a *landscape change*. The outcome of the shift proposed by Stahel would therefore concern both the socio-technical regime and the socio-technical landscape in Geels and Shot's multilevel perspective (Geels and Schot, 2007).

4 Mapping the CE Landscape

The evolution from the earliest concepts related to sustainability to the concept of Circular Economy is visible in different dimensions: one dimension concerned the expansion from the initial focus on critical sectors (packaging and automotive) to all the areas related to industrial production (Howard et al., 2022; Emf, 2021); another dimension concerned the expansion from the initial focus on extended produced responsibility to a redesign approach (Ceschin and Gaziulusoy, 2019; Simeone et al., 2019), which could facilitate maintenance, repair, remanufacturing, and material recycling; a third dimension concerned the extension of the focus from products to their use or functionality (Stahel, 2010), which was focusing on infrastructures (services) and socio-cultural changes that would make circular economy possible.

The concept of CE was probably the catalyst that inspired a number of initiatives in the last decades, which populated a panorama of niche innovations (from products to services), infrastructural changes (local, national, and international legislation), and socio-cultural changes. Among policy innovations, the European Green Deal proposed a large socio-cultural change that was expressed through a roadmap and, more specifically, the European Circular Economy Action Plan (EC, 2020) was setting the regime conditions for the CE in different sectors. At the same time, the Green Deal itself is the sign of a large landscape transition toward an institutional culture in which the question of sustainability becomes crucial to any development of society, economy, and technology. The result of the change that occurred in the last ten years is in the wide range of initiatives undertaken in every sector, with the aim of exploring the multiple dimensions that the concept of CE implies.

The various dimensions described so far propose a good basis for mapping CE initiatives in a specific local, regional, national or international context. The authors of this paper, however, tried to go one step further and map a specific landscape, i.e., a specific national context, to address two research questions:

- How can we recognize and qualify the different dimensions of changes toward CE in a real context? This question implies an observation of how the change related to CE is actually happening in this context. This observation, in turn, is particularly relevant to inspire future initiatives and could also address a second important research question.
- How can a change toward CE economy be designed? This question is particularly relevant from two perspectives. From a perspective focusing on socio-technical systems, it is important to see to what extent the systemic

change related to circular economy can be addressed, i.e., can be designed. This means taking into account how favorable developments can be included to enhance the horizontal scalability of relevant niches or the vertical scalability (i.e., the institutionalization) of some preferable changes. From a designer's perspective, the question should introduce a reflection on the role that design can have in such a change and on the design competencies that can support it.

5 The Focus on Portugal as the Main Case

The national context analyzed in this paper is Portugal. The choice has been suggested by the location of one of the authors, and therefore, by the possibility to access relevant empirical material. It is worth noticing that the information in this context is by no means complete, as cases and initiatives on CE are continuously developing. However, the choice of a specific set of data related to this context makes it possible to refine an approach for reading and interpreting change in light of CE principles and perspectives.

Since 2014, the EU has promoted CE strategies through various projects and initiatives, the most recent and robust of which is the 2018 EU Circular Economy Package (Bourguignon, 2018). The EU also promoted an action plan for circular economy, aimed at ensuring a regulatory framework for the mobilization of economic operators and society at large in promoting initiatives toward waste treatment strategies oriented toward circularity. The EU action plan focused on actions not only at the EU level, but also highlighted the need for long-term involvement at all levels, including states, regions, and cities (EC, 2015).

In 2017, Portugal approved the Action Plan for CE, The Portuguese action plan ("Leading the transition", 2017) sets the action at the macro, meso (sectorial), and micro (regional/local) levels.

The macro level includes measures at the strategic level, which focus on seven different aspects, such as extended producer responsibility, education, food waste reduction, resource regeneration, and research and innovation. The implementation of these initiatives was based on political instruments such as green taxation or voluntary agreements; at the sectoral and regional levels, other incentives were put in place, i.e., funding, that could support the development of solutions, e.g., fundo ambiental, fundo para a inovação, tecnologia e economia circular, Portugal 2020 funding program.

The meso level sets the agendas for action, concerning the efficiency of material productivity and public procurement. At this level, the action plan indicates key sectors of the Portuguese economy such as tourism, textile, shoe industry, retail, and resource intensive sectors (construction). Finally, the micro level focuses on actions at the local level, setting expectations for initiatives from companies, local administrations or other local stakeholders (ETC/WMGE, 2019); at this level, the agendas focus on developing circular cities, organizations/companies, and promote industrial symbiosis. The action plan also triggered reflections and studies on *industrial symbiosis* as a strategic approach to make sure that waste from a production/consumption chain could be reintroduced into another productive chain (BCSD, 2018).

In early November 2022, the evaluation of the implementation of the plan and the public consultancy of research results was undertaken to support the development of the second Action Plan for CE in the country (LNEG, 2022). Seven dimensions for action were proposed, identifying driving forces and barriers to further infuse CE in culture, investment, technology and infrastructure, policy and regulation, stakeholders' network as well as institutional and organizational dimensions. Despite being identified as a leading country in terms of circularity initiatives (Winans et al., 2021), major barriers regarding design infusion were identified, especially regarding the "lack of infrastructures that enable circularity"; "lack of investment that still exists in ecodesign", and "products that are not prepared for circularity" (LNEG, 2022). As mentioned by the founder of the Circular Economy Portugal Association, and consultancy advisor for the government, in conversation with one of the co-authors: "circular design needs to be more emphasized in this second action plan" (mentioned during a project meeting with the co-author).

Circular Economy initiatives in Portugal.

To help us identify the initiatives, this paper includes the analysis of a series of CE initiatives developed in Portugal based on the information collected from a large database issued by the Circular Economy Club (CEC)—an international nonprofit network, including professionals and organizations with over 280 CEC local clubs in 140 countries. Portugal has been active since 2020 and has currently 7 active clubs scattered throughout the country namely in Porto, Braga, Felgueiras (northern region), Lisbon and Almada (center), and Faro (southern region).

The study of the Circular Economy Club refers to a mapping week in February 2018, which engaged 2100 participants from 40 countries through 65 workshops. The workshops developed a database of 3,000 initiatives related to CE, but the database was further developed after the mapping week. Operating from this database, the authors identified 98 initiatives related to Portugal.

The database included different aspects collected by the participants of the workshops of the Circular Economy Clubs. One of them consisted in the identification of (1) six CE strategies namely, design, resources, business models (BM), product life extension (PLE), waste as a resource, and other strategies, and (2) six main sectors in which innovation appeared such as cities (including buildings, infrastructure, and mobility), consumer products and electronics, fashion and textiles, food and beverage, manufacturing, and other sectors (including financial services, healthcare, education, and tourism). Finally (3), the database identified the type of organization leading the CE initiatives: Multinational Corporate, National Corporate, SME, Startup, Private sector, Education, Non-profit, Government, Support (investment, consulting, media, etc.), and others.

5.1 Preliminary Results of Existing Data and Categories

The analysis of preliminary results shows that Portugal appears at the 12th place in the ranking (3,24%) of the CE initiatives mapped by the CEC around the world, with 50 participating countries. The country is preceded by the UK (with 13% of representation), Sweden, Germany, Canada, South Africa (between 5–10% representation); and Taiwan, Austria, Denmark, Spain, USA, and Chile (between 3,5–5% representation).

In comparison to the rest of the world, Portugal has identified more initiatives in specific key sectors of its economy, namely in Fashion and Food/Beverages (with approximately +5% in each organizational sector) (Fig. 2). Organizational sectors with less representation are Manufacturing (−6%) and Cities (−4%).

Additionally, when analyzing the type of organizations involved in the CE initiatives, a major difference in representativeness of the private sector (−19%) and SMEs (−16%) was identified, when compared to other countries. In contrast, a higher number of Non-profit (+8%), National Corporate (+9% approximately), Startups (+11%), CE initiatives in the Food/Beverage (for Non-profit and National corporate), and Fashion sectors (for the Startups) were identified in the database (Fig. 3).

Finally, when compared to other countries, the primary CE strategies in Portugal are more focused on designing new business models (Fig. 4). This type of strategy is mostly used in the consumer products/electronics sector.

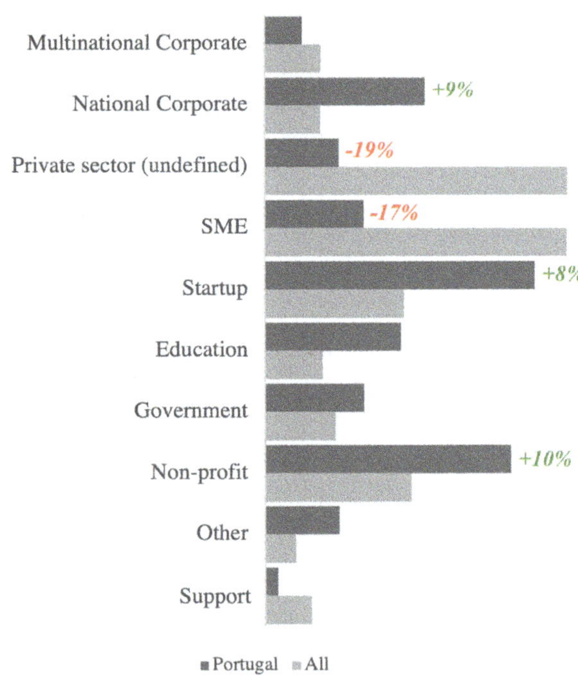

Fig. 3 CE initiatives by actors

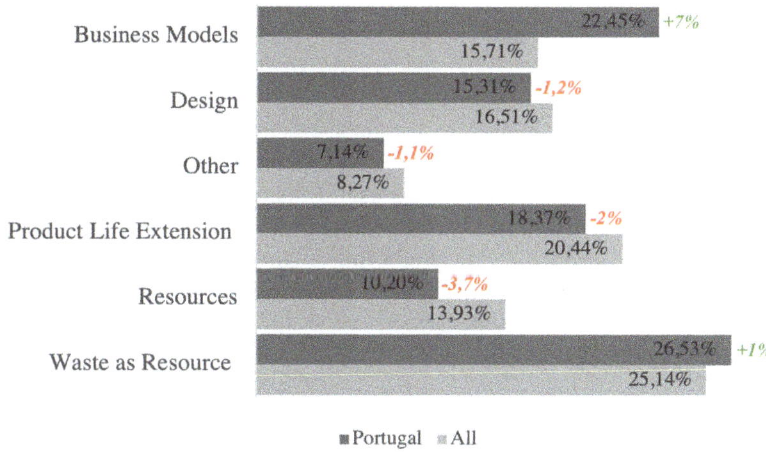

Fig. 4 CE initiatives by CE primary strategy

When highlighting the analysis of the CE initiatives collected in Portugal specifically, the main organizational sectors involved in such initiatives are food/beverages (22%). A large part (21%) of the initiatives were also concerned with urban components (Cities—buildings, infrastructure, mobility, logistics, and resource management) and Consumer Products/Electronics (21%). The Fashion sector is also included with 17% of initiatives and manufacturing has the lowest representation (7% of cases).

Second, up to one third of the organizations/entities leading CE activities are Startups (~23%) and Non-profit organizations (~20%). This data indicates that the institutionalization of CE may be still at an early stage in the country (Salamzadeh and Kesim, 2015), as startups are newly born companies that often struggle to consolidate their business in the mainstream market. The ideas and prototypes are still at the testing stages, often construed with high uncertainty of success; the high participation of non-profit organizations in CE initiatives can also indicate that CE may yet not be profitable and still be limited to small niches, with limited potential for scaling out (replicating) (Moore and Riddell, 2015) such initiatives in other contexts. Some initiatives do rely partially on voluntary citizens' movement and participation, finding alternative ways of doing/living, which are complementary (but sometimes not compatible) with market growth (e.g., permaculture).

Third, the most recurrent CE strategy used in Portugal, according to this database, concerns the reuse of Waste as a Resource (~27% of the initiatives concern this strategy); another substantial quote (22%) of initiatives concerns the development of New Business Models. Product Life Extension and Product Redesign were also recurring with a lower frequency (18% and 15%, respectively). When crossing the different categories as seen in Fig. 5, Waste as Resource CE strategy is embedded in initiatives identified throughout the different sectors. Other strategies, however, seem to be more secluded to specific sectors. For example,

"Design" is used mostly in the fashion and manufacturing sectors. "PLE" is mostly used in Consumer Products and food/beverages sectors. The "Business Models" strategy is identified in four sectors—but is more intensively used in the Consumer Products sector.

In summary, the report related to the database includes some categorizations that are useful to read the change CE is defining in societies. The interpretation of the model, however, does not provide useful insights to understand the transitional value of CE initiatives, i.e., to understand how such initiatives are triggering change at the micro level (e.g., by promoting a single technical initiative), or at the meso level (e.g., by generating a system based on a series of services or social interactions between different stakeholders) or at the macro level (i.e., at the level of major socio-technical changes).

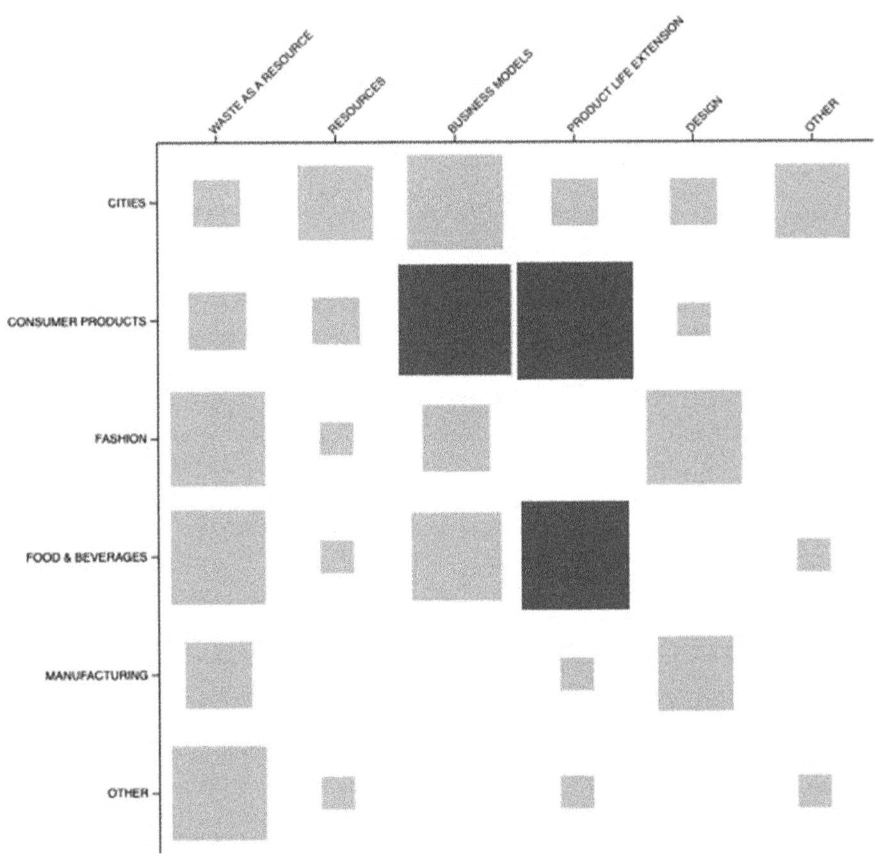

Fig. 5 CE primary strategy and organizational sectors

6 Some Considerations About the Database

While the database provides a good *logical geography* to navigate and understand the development of CE initiatives in Portugal, it does not provide a key to understanding the potential impact of such initiatives. On its own, this database does not help understand the evolution of CE in Portugal.

A first consideration should concern the scale of innovation, which in turn is connected to the complexity of the initiative.

Because of its comprehensiveness, the strategy framework proposed by Alaerts et al. (2019) (see Table 1) can be used as a first filter to interpret the database of initiatives in Portugal. The authors of this paper tried to map the criteria proposed in the framework of the initiatives in the database. While the mapping is purely based on the authors' interpretation of the database, it gives a first glance at the distribution of the initiatives along different levels of complexity. It is worth mentioning, though, that some initiatives could not be included in the framework: the database indeed, reported initiatives related to the creation of (a) educational initiatives, aimed at creating capacities (i.e., diffuse knowledge among new generations) and (b) policy or government actions. For this reason, two new categories have been added to Alaerts et al.' categorization: education and policies and government actions.

Figure 6 provides a visual overview of the strategies, as they have been identified in the database. The figure shows how the most recurrent strategies concern the activities of recycling and repurposing, with a lower incidence of initiatives aimed at recovering energy from discarded materials, which is destroying most of the properties, knowledge, and characteristics of products. Interestingly, many initiatives refer to systemic aspects, i.e., to the higher part of Table 1. Those are initiatives to exchange second-hand products, sometimes creating marketplaces, or to use less resources or re-focusing on functions, thus shifting to different production and consumption configurations.

7 Beyond the Data

In order to address the first research question proposed in this paper (*how can we recognize and qualify the different dimensions of changes in a real context?*) a further step is needed: it is important to *enter* into the data and investigate the nature of the initiatives. For this purpose, the authors collected further information about the initiatives, using the links and documentation reported in the database, and interrogated the database according to different perspectives, using additional labels that open to different categorizations.

One of the most meaningful characterizations concerns the focus of the initiative: the landscape observed through the database included initiatives of very different nature, from intervention on buildings to education, from the redesign of a product to the definition of marketplaces for second-hand products and to

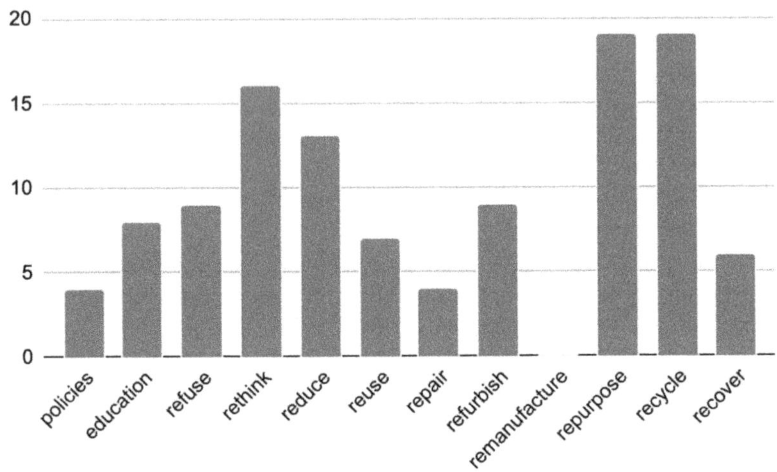

Fig. 6 CE strategies in Portugal

the definition of policies to support circular economy. The authors used different labels, possibly attributing more than one label to the same initiatives.

The foci defined in Table 2 could be further grouped:

- A group of initiatives that focuses on the physical or molecular manipulation of materials (e.g., recycling) or resources (e.g., water saving). Those initiatives are mostly related to level C in the classification proposed by Alaerts et al. (2019). This level includes innovation related to buildings, products, and resource management. Most of those initiatives rely on changes based on technologies that make it possible to reuse materials or resources (from cigarette butts to water or codfish spines). The database does not provide clear clues on the reason why those initiatives are more diffused than others at level A or B; a hypothesis could be that those initiatives can be better controlled by one or few actors (those who handle the technology), with reduced involvement of other actors (such as household, communities or schools), that usually have a role in the collection of discarded products (such as electronics or food). Since they do not require major behavioral change, those initiatives may be easy to establish and consolidate.
- A group of initiatives that works on the creation of exchange platforms, marketplaces, special events or services; these are initiatives that create the conditions for reuse, refurbishment, repurposing or repair (e.g., material repositories, children's toy sharing), and for rethinking the way products are used (e.g., car sharing). These initiatives are related to levels A and B in Alaerts' classification. On the other hand, these initiatives could be defined as *infrastructuring* initiatives, as they generate *regime conditions* for activating flows for the exchange or recirculation of products, components or materials.

Table 2 Main focus of the initiatives

Focus	Description
Building	Initiatives that concern the building sector
Education	Initiatives for schools or training actions or actions that increase the awareness of possible circular flows in a community
Infrastructuring	Initiatives that generate opportunities for smarter use of products and services or that support the extension of products' lifespan. Within infrastructuring, further subcategories have been identified: Platforms: creation of opportunities for material reuse, swap marketplaces, multiple use of products (sharing) Events: organization of fairs, periodic markets, and demonstration events that increase citizens' awareness of the opportunities for circular actions Services: modification of existing services to optimize the use of resources (e.g., reducing packaging)
Policies or government actions	Policies: policy actions aimed at creating roadmaps or subsidizing circular economy initiatives
Product	Initiatives that focus on products, to increase their lifespan (repairing, reusing, remanufacturing) or to optimize the use of materials
Resources	Initiatives for the reutilization of material for the optimization of resources (e.g., waste into energy or water)

– A third group of initiatives concerns initiatives aimed at generating capacities at the systemic level, by proposing specific policies or by activating education initiatives that tend to trigger cultural change.

It is worth noticing that this grouping provides some clues to compare the observed innovation landscape to the multilevel structure proposed by Geels and Schot (2007):

1. Rather than proposing any systemic action, the initiatives in the first group focus on specific materials or products or resources, thus providing solutions that concern a specific niche, with its problematic implications. Those initiatives often imply the use of specialized technical knowledge (from architects to chemical engineers), and they represent a good opportunity to advance knowledge on specific technical problems. For example, the "redbridge school" project uses wood and other renewable or low-impact materials to build the structures and facades of the school. In the fashion sector, a well-known shoe brand "As Portuguesas" uses residues of the cork from the stopper industry as the main raw material to create the shoe soles—thus using existing materials to generate new value via product design. Finally, looking at resources, other examples can be identified, to recycle or recover the material, namely, the initiative developed by EPAL in Lisbon, that uses treated wastewater to be used in public gardens

maintenance, or cleaning public facilities and goods such as metro stations or buses.

2. The infrastructuring initiatives concern the creation of swap markets or (sometimes local) marketplaces or platforms, which activate new practices, thus new socio-technical routines, that can change the *regime* conditions in the Portuguese context. It is worth noticing, however, that a regime change, as described by Geels and Schot (2007), includes changes in the whole technical and institutional infrastructure of a context, which cannot be observed from the initiatives in this database, possibly because such change cannot be observed from any single initiative but is rather a systemic emergence. This group of initiatives mostly involves actors with different skills, including design, communication, or social skills. The infrastructuring initiatives in the dataset at hand appear to have a wider spectrum of intervention types—from creating platforms to events, services, and policies, at the local and national levels. For example, initiatives related to "events" are mostly focused on organizing markets for reselling or exchanging second-hand items; and/or service events to repair appliances/objects. Services include initiatives related to "smarter product use and manufacture", supporting citizens to extend the products' life, namely for system needs related to mobility (car and bike sharing), as well as with education (centers for collecting and sharing schoolbooks).
3. The generation of platforms is also a strong mechanism to support infrastructuring toward CE. Platforms can be defined as a "set of actors, resources, business and operational rules and setting that converge for value creation" (Breidbach et al., 2014).
4. Although platforms are often digitally enabled (e.g., Booking Drive, an online shop to rent all kinds of equipment), the database also provides a set of examples of alternative mechanisms to adjust and develop the conditions for smarter product/service use and needs fulfillment, for example, by creating alternative channels for zero waste of food and clothes (e.g., Fruta Feia that works with wasted/ugly fruits/veggies, which do not fit market requirement for retails, are sold to end customers based on a monthly fee); or chain of shops to sell/buy or exchange second-hand goods/furniture.
5. Finally, the third group of initiatives aims at a broader change in the socio-technical system. If effective, those initiatives will impact the institutional system and the cultural landscape of the country, therefore, they could be compared to changes in what Geels and Schot define as the socio-technical landscape. Obviously, the database only offers a perspective on the potential landscape change, rather than a real view on something really happening. Policy initiatives include the definition of roadmaps or action plans for the public support of CE at the municipal as well as the national level. The potential of such an initiative is in the capability to create seamless support to an otherwise fragmentary landscape of small changes that can be observed through the database. Education is also an important vector for capacity building, that spreads throughout all the levels of complexity related to CE, including:

- Initiatives to promote CE in schools and public spaces, creating education programs, courses and providing training to change existing practices (level A in Table 1), Examples include “escolas circulares” (circular schools), or “vegetable garden at your door” that provides citizens with garden plots to practice organic agriculture and receive training in farming and composting;
- Educational initiatives that address repair or refurbishment (level B in Table 1), such as the RLAB, a knowledge laboratory to learn to recover electric/electronic equipment;
- Educational activities that support the reduction of centralized treatment of organic waste or workshops about composting (level C in Table 1). The long-term impact of these initiatives in consumer behavior and market changes, through time, is, however, still a topic that requires further inquiry.

The cases mentioned in this section can be consulted in the file annexed to this chapter. They are also available and updated regularly in the Circular Economy Club open-access repositories (https://www.circulareconomyclub.com/gd-home/cec-global-database/).

8 Designing Circular Economy

The considerations proposed in the previous sessions provide clues to address the second research question: whether (and to what extent) circular economy can be designed. The systemic nature of CE implies that several actors and several competencies must be involved in the process of change. Technical competencies are crucial for the process of change, especially regarding the manipulation of the physical and chemical composition of materials and products; however, as seen in the observation of the database, other competencies are also crucial, because they generate societal conditions for change, in organizational, social, and cultural terms. Many actors are able to contribute to such change, and therefore, according to Simon’s (Simon, 1969) definition, many actors are *designing* this change according to their own competencies. Those that, according to Manzini (Manzini, 2015), could be defined as expert designers, can contribute to the initiatives in the various groups identified in the database with different competences.

The initiatives focusing on products and materials require designers to redesign products, reuse materials from previous cycles or facilitate product disassembly or modularity. This area is not new for designers, as it was perceived as a critical challenge to approach Extended Producer Responsibility, and therefore, developed with several contributions for several decades (Ceschin and Gaziulusoy, 2019; Lewis et al., 2001). The initiatives focusing on infrastructuring are urging designers to find solutions that challenge the existing socio-technical system. The three main areas defined in Table 2 are platforms, events, and services.

Platforms generate opportunities for interaction among different stakeholders: they could consist of online or local swap markets that recirculate used products or materials or even useful knowledge. The level of complexity can be quite different.

Some of the initiatives, such as *Fashion* Revolution are relying on existing social networks for the interaction among the users, while they provide useful information in their website. Other initiatives are in fact promoting the exchange of material (e.g., "Repositório de materiais"—Repository for materials), but because of the nature of the material exchanged (building and construction material), need to focus on specific geographical areas, possibly planning to scale out by nodes, i.e., generating local material exchange networks in different cities. Finally, other platform initiatives are based on an accurate design of the interaction between different stakeholders: *Refood*, for instance, has created an ecosystem that includes supermarkets, restaurants, food producers, voluntaries, and other actors to produce meals for persons in need and institutions. The design of a platform requires a good understanding of systemic aspects, including the interaction between different stakeholders and their mutual negotiation. As seen in the database, platforms can emerge from social interaction within existing challenges, but they can also have a higher level of complexity, in which the interaction, i.e., the way value is exchanged among the various stakeholders, needs to be accurately organized. With respect to this, the effort of product/service design should therefore focus on understanding the complexity of such systemic aspects: designers should focus on generating appropriate frameworks for action, rather than specific solutions (EC, 2020).

Events are temporary opportunities for product exchanges (e.g., flea markets) or for getting knowledge on how to repair or extend products' life. The social nature of such events makes it possible to organize them through the most common social channels, or by using existing networks that have been created around fablabs or other active organizations. They are obviously local events, even though initiatives like bike repair or repair cafés are increasingly replicated in several locations (e.g., Repair café Lisboa). Although some of those events are the result of a *diffuse* design capability (Manzini, 2015), the way in which those initiatives are replicated or scaled out is an interesting design problem, as they may imply a capability to understand contexts, engage relevant stakeholders and facilitate social aggregation (Morelli, 2014; Concilio et al., 2013; Morelli, 2015).

Because of their interactive nature, platforms and events can be often considered as **services**, i.e., structured solutions that mobilize a number of stakeholders and technologies to address specific needs related to the utilization of products, or create specific value for different beneficiaries. The database, however, also includes initiatives consisting of specific solutions that support the reduction in resources or material use (e.g., the suppression of packaging) and or the rental of goods that are not frequently used, such as luggage or baby equipment.

The initiatives aimed at changing the socio-technical landscape to support CE include educational initiatives and policy or government actions. While designers have been involved in different educational initiatives, the participation of design in policy actions or to institutional structures is relatively new and opens new perspectives for designers. The initiatives in the database define roadmaps or action plans for the intervention of public institutions in relevant environmental problems. Designers are increasingly involved in this strategic process (Whicher et al.,

2018). The definition of missions is becoming part of the strategic planning at different institutional levels (Mazzucato, 2018; Miedzinski et al., 2019) and the European Green Deal EC (2019) is a relevant outcome of this approach. The definition of research and innovation programs at the EU level clearly recognizes the role of design in coordinating and synergizing different elements of innovation through visions and consistent scenarios, but also in translating the elements of such visions into concrete projects. While it is not possible to identify this role in any specific initiative in the database, it is possible to figure out the role design can have when translating the wide program of the Portuguese roadmap for the circular economy into initiatives at the local level, promoting operative action and participatory initiatives to support sustainable plans at the municipal level.

9 Conclusions

CE is clearly calling for a broad systemic view, which links initiatives in niche sectors or on specific products or materials to larger infrastructural changes and to systemic initiatives, which orient the socio-technical landscape toward circular flows. This paper tried to read a "real" context, as described by a large database of initiatives, through the lens of multilevel innovation, as proposed by Geels and Schot; the idea, in using this filter, was to make sense of the data in the database. Geels and Schot's perspective is not completely adequate to interpret the database (or at least the database does not offer enough information to relate the situation to Geels and Schot's perspective), however, it gives the spark to an analysis of a context that would otherwise seem too articulated and complex. It also gives a chance to highlight some initiatives that go beyond the traditional action of designers, following the evolution of designers' professional domain, from products to services, and more recently to systemic and institutional perspectives.

The challenge that CE proposes to designers is to define their actions through a full palette of capabilities (Morelli et al., 2020). While in the early phases of the debate on sustainability, designers were working around products and their material components, they are now asked to use their capability to activate social processes, understand and support mechanisms of scalability, work across different levels of complexity, and translate broad visions about the future of this society into operative projects and organizational structures.

It is important to take into account that the database analyzed in this paper is to be considered as a sample, rather than the complete description of the innovation landscape around CE in Portugal. The whole landscape is continuously changing, and new initiatives are being proposed. The database only represents a ground for discussion on (a) how a myriad of innovation phenomena at different scales is shaping broad changes; (b) how this change can somehow be oriented through a design action and; (c) what kind of challenges designers are facing, when entering in the mechanisms of change activated by CE.

References

Alaerts L, Van Acker K, Rousseau S, De Jaeger S, Moraga G, Dewulf Jo et al (2019) Towards a circular economy monitor for Flanders: a conceptual basis. Conclusions of Stakeholder Workshop, June 27, 2018. Pub. N. 4. https://vlaanderen-circulair.be/src/Frontend/Files/userfiles/files/Towards%20a%20CE%20monitor%20for%20Flanders.pdf

Ayres RU, Simonis UE (1994) Industrial metabolism: restructuring for sustainable development. United Nations University Press, Tokyo, New York

BCSD (2018) Circular synergies, a challenge for Portugal. BCSD Portugal–Business Council for Sustainable Development. https://bcsdportugal.org/wp-content/uploads/2018/08/BCSD-Portugal_Circular-Synergies_2018.pdf

Bourguignon D (2018) Circular economy package four legislative proposals on waste. Briefing. EU legislation in progress. Environment, Public Health and Food Safety (ENVI)

Braungart M, McDonough W (2008) Cradle to cradle: remaking the way we make things. Vintage, London

Breidbach CF, Brodie R, Hollebeek L (2014) Beyond virtuality: from engagement platforms to engagement ecosystems. Manag Serv Qual 24(6):592−611. https://doi.org/10.1108/MSQ-08-2013-0158

Cahill R, Grimes SM, Wilson DC (2011) Extended producer responsibility for packaging wastes and WEEE-a comparison of implementation and the role of local authorities across Europe. Waste Manage Res 29(5):455–479. https://doi.org/10.1177/0734242X10379455

Campbell-Johnston K, ten Cate J, Elfering-Petrovic M, Gupta J (2019) City level circular transitions: barriers and limits in Amsterdam, Utrecht and the Hague. J Clean Prod 235:1232−1239 https://doi.org/10.1016/j.jclepro.2019.06.106

Ceschin F, Gaziulusoy İ (2019) Design for sustainability: a multi-level framework from products to socio-technical systems, 1st edn. Routledge

Concilio G, Deserti A, Molinari F, Puerari E, Rizzo F (2013) Scalability methodology [Project Deliverable]. Politecnico di Milano, MyNeighborhood Project D1.5, Milano, Lisbon

EC (2015) Communication from the commission to the European parliament, the council, the European economic and social committee and the committee of the regions: closing the loop-an EU action plan for the circular economy. 2.12.2015. European Commission, Brussels, p 21 COM/2015/0614 final

EC (2019) Communication from the commission to the European parliament, the European council, the council, the European economic and social committee and the committee of the regions. European Green Deal. 11.12.2019. European Commission, Brussels

EC (2020) Circular economy action plan. European Union

Emf S, McKinsey (2021) Growth within: a circular economy vision for a competitive europe [Report]. Ellen Macarthur Foundation

ETC/WMGE (2019) Resource efficiency and circular economy in Europe–even more from less. An overview of policies, approaches and targets of Portugal in 2018. Eionet report. 4/2019. European Environment Agency. https://www.eionet.europa.eu/etcs/etc-ce/products/country-factsheets/b-country-profile-portugal_finalised.pdf

Frosch RA, Gallopoulos NE (1989) Strategies for manufacturing. Sci Am 261(3):144–153. https://doi.org/10.1038/scientificamerican0989-144

Frosch RA (1992) Industrial ecology: a philosophical introduction. Proc Natl Acad Sci 89(3):800−803. https://doi.org/10.1073/pnas.89.3.800

Geels FW, Schot J (2007) Typology of sociotechnical transition pathways. Res Policy 36:399−417. https://doi.org/10.1016/j.respol.2007.01.003

Gertsakis J, Morelli N, Ryan C (1998) Return to sender; An introduction to extended producer responsibility. National Centre for Design at RMIT University, Melbourne, Australia

Gertzakis J, Morelli N, Ryan CJ (2002) Industrial ecology and extended producer responsibility. In: Handbook of industrial ecology. Edward Elgar Publishing, London, pp 521–529

Giarini O, Stahel WR (1986) The functional economy, cultural and organizational change. Sci & Public Policy 13(4). http://www.product-life.org/en/archive/the-functional-economy-cultural-and-organizational-change
Heiskanen E (1996) Conditions for product life extension. National consumer research centre working papers, University of Helsinki, Helsinki
Howard M, Yan X, Mustafee N, Charnley F, Böhm S, Pascucci S (2022) Going beyond waste reduction: exploring tools and methods for circular economy adoption in small-medium enterprises. Resour Conserv Recycl 182:106345. https://doi.org/10.1016/j.resconrec.2022.106345
Leading the Transition. Action Plan for Circular Economy in Portugal: 2017–2020 (2017) Republica Portuguesa. p 62. https://circulareconomy.europa.eu/platform/sites/default/files/strategy_-_portuguese_action_plan_paec_en_version_3.pdf
Lewis H, Gertsakis J, Grant T, Morelli N, Sweatman A (2001) Design and environment: a global guide to designing greener goods. Greenleaf Publishing, London
LNEG (2022) Situation of the circular economy in Portugal: barriers and driving forces for the acceleration of a circular economy in Portugal. Retrieved from https://videocast.fccn.pt/live/id_tecnico_ulisboa/nlirut
MacArthur E (2013) Towards the circular economy. J Ind Ecol 2(1):23–44
Manzini E (2015) Design, when everybody designs (Trans: R. Coad). Cambridge, Massachusetts, London, England: MIT Press
Mazzucato M (2018) Mission-oriented research & innovation in the European union. A problem solving approach to fuel innovation-led growth. European Commission, Brussels. https://ec.europa.eu/info/sites/info/files/mazzucato_report_2018.pdf
Melles G (2021) Figuring the transition from circular economy to circular society in Australia
Miedzinski M, Mazzucato M, Ekins P (2019) A framework for mission-oriented innovation policy roadmapping for the SDGs: the case of plastic-free oceans [Working Paper]. UCL Institute for Innovation and Public Purpose, London, p 61 2019–03
Moore ML, Riddell DJ (2015) Scaling out, scaling up, scaling deep: advancing systemic social innovation and the learning processes to support it. J Corp Citiz 58:67–84. https://doi.org/10.9774/GLEAF.4700.2015.ju.00009
Morelli N, De Götzen A, Simeone L (2020) Service design capabilities. Springer, Cham
Morelli N (2014) Challenges in designing and scaling up community services. In: Sangiorgi D, Jugninger S ServDes2014. Lancaster University, Lancaster
Morelli N (2015) Designing for few, scaling up for many. In: Valmenburg R, Dekkens C, Sluijs J Participatory innovation conference (PIN-C). The Hague University of Applied science, Syddansk Universitet, The Netherlands, The Hague. https://pin-c.sdu.dk/assets/designing-for-few-and-scaling-up-for-many---nicola-morelli---209-216---pinc-2015.pdf
Salamzadeh A, Kesim HK (2015) Startup companies: life cycle and challenges. In: 4th international conference on employment, education and entrepreneurship (EEE). Belgrade, Serbia
Simeone L, van Dam K, Morelli N (2019) A preliminary review of the concept of circular economy in design research [Article in proceeding].
Simon HA (1969) The sciences of the artificial. MIT Press, Cambridge
Stahel WR (1994) The utilisation-focused service economy: resource efficiency and product-life extension. In: Alenby Braden RDJ (ed) The greening of industrial ecosystems. National Academy Press, Washington D.C. https://nap.nationalacademies.org/read/2129/chapter/17
Stahel WR (1997) The functional economy: cultural and organisational change. In: The industrial green game. National Academy Press, Washington, DC, pp 91–100. https://nap.nationalacademies.org/read/4982/chapter/7

Stahel WR (2010) The performance economy. Palgrave Macmillan, Basingstoke, England, New York

Whicher A, Harris C, Beverley K, Swiatek P (2018) Design for circular economy: developing an action plan for Scotland. J Clean Prod 172:3237–3248. https://doi.org/10.1016/j.jclepro.2017.11.009

Winans KS, Mkrtchyan I, Gonçalves JM (2021) Approaches to the circular economy in Armenia and Portugal: an overview. In: Circular economy and sustainability: volume 1: management and policy, vol 1. pp 651–661. https://doi.org/10.1016/B978-0-12-819817-9.00003-X

5 Circular Packaging in the Cosmetics Industry—A Systematic Review on Challenges and the Current State of Sustainable Strategies and Solutions

Christoph Scope, Nikolas Neumann, Christian Wölfel, and Linda Kästner

1 Introduction

The greatest challenges for humankind in twenty-first century include avoiding carbon emissions, protecting habitats, and conserving resources. However, the use and purchase of consumer goods harm these goals. Plastic packaging in particular has a major impact. The largest share of plastics used throughout the European Union (EU), 39.1%, is used for packaging. By comparison, the next largest application area, building and construction, is responsible for 21.3% of plastics (Plastics Europe and EPRO 2022). In this chapter, we review existing efficiency, sufficiency, and consistency strategies to enable sustainable packaging in the cosmetics industry with its exacerbated requirements and identify current strengths and weaknesses of proposals to date. We argue that far more attention to upstream innovation rather than end-of-pipe recycling solutions is required.

C. Scope · N. Neumann · C. Wölfel (✉) · L. Kästner
Technische Universität Dresden, 01062 Dresden, Germany
e-mail: christian.woelfel@tu-dresden.de

C. Scope
e-mail: christoph.scope@tu-dresden.de

N. Neumann
e-mail: nikolas.neumann@tu-dresden.de

L. Kästner
e-mail: linda.kaestner@mailbox.tu-dresden.de

© The Author(s), under exclusive license to Springer Nature Singapore Pte Ltd. 2024

G. B. Melles and C. Wölfel (eds.), *Design for a Sustainable Circular Economy*, Design Science and Innovation, https://doi.org/10.1007/978-981-99-7532-7_5

1.1 SDG 12, Sustainability Transition and the Circular Economy

The United Nations (UN) General Assembly has adopted 17 Sustainable Development Goals (SDGs) in 2015, number 12 of which concern sustainable consumption and production (UN Department of Economic and Social Affairs (UNESA) 2022). The goals of SDG 12 include preventing waste and encouraging more sustainable patterns of consumption, for which the concept of the circular economy (CE) is being pursued in particular. The positive contributions of circular strategies on SDGs have been examined by many authors (Geissdoerfer et al. 2017; Rodriguez-Anton et al. 2019; Schroeder et al. 2019).

It is argued that CE concepts may entail "the potential to induce transformative sustainability change" (Reike et al. 2018, p. 247). For this reason, science and policy encourage closing materials loops and avoiding unsustainable disposal. A real sustainable transition needs change at multiple levels (Geels 2011; Melles et al. 2022a, b), especially given that global circularity has fallen from 9.1% in 2018 to 7.2% in 2023 (Circle Economy 2023).

At the macro level, there is a slowly changing socio-technical regime that is hesitant to radical changes, being characterized not only by policies and norms, but also by common practices and conventions. On a micro level, niche innovations emerge and design can act as an enabler to provide novel (niche) solutions. Designers could therefore influence the socio-technical regime (Melles et al. 2022a, b).

Accordingly, there is a call for design to provide new sustainable solutions. The design has the means to contribute to a number of SDGs, e.g., by enabling participatory development (Velenturf and Purnell 2021). With regard to SDG 12, the design discipline can promote a sustainable CE with drastically reduced resource consumption, closed material loops, and low/zero carbon footprints. Strategies should be socially responsible and foster the overarching principles of efficiency, consistency, and sufficiency (Schuster and Throl 2021).

As a rule of thumb, strategies that move as far away as possible from a linear economy and are as close as possible to the idea of a CE are preferable. Refuse and reuse are, therefore, superior to recycling strategies in terms of their circularity. The validity of this sustainability hierarchy is verified in concrete individual cases with the aid of an life-cycle-assessment (LCA) assessment (Circular Economy Initiative Deutschland 2021; Potting et al. 2017).

1.2 The Circular Economy, Packaging, and Cosmetics Challenge

Although refusing packaging is a long-term strategy with a high impact on sufficiency, packaging is needed to serve the core functions of protecting the product. Thus, there is great potential for savings in packaging in terms of material efficiency and energy waste. For example, in Germany, not even 50% of plastics in large parts packaging are recycled. Although recent regulations demand increased recycling quotas in packaging (Directive 94/62/EC 1994; VerpackG 2021), it is not yet clear how these quotas can be achieved based on the current state of the art in recycling technologies, business models, packaging designs, and other constraints. Part of that problem is that recycling plastic waste is still very costly (Schuster and Throl 2021).

In the cosmetics sector, all those challenges are exacerbated by, e.g., special hygienic requirements and social networks that motivate increased consumption (Dinh and Lee 2021). Despite these problems, there are already more sustainable packaging alternatives on the market. But the reasons why customers do not opt for these 'green' products are, again, manifold, such as shape, functional properties, or personal preferences. As a product's design and conception phase determines the entire product life cycle at an early stage, packaging designers and managers should include the purchase decision processes in their sustainability and CE strategies.

To help practitioners compare products and packaging, an "integrated circularity-sustainability assessment" (Hatzfeld et al. 2022, p. 2) is aimed for. But overall, research on the interconnection of sustainable and circular packaging design, particularly for cosmetics and hygiene products, seems to be rare. In order to address this research gap, we aim to evaluate the current state of research by means of a systematic literature review on CE strategies for cosmetics and hygiene products. We assess if and how the circular strategies meet sustainable aspects. We follow the triple-bottom-line definition of sustainability (Elkington 2018; Finkbeiner et al. 2010) and on findings that are directly related to the packaging design for circular solutions in the cosmetics and hygiene sector. The review is guided by the following questions:

- What is the functional role of packaging for cosmetics and hygiene products?
- Which sustainable and circular packaging design approaches exist for cosmetics and hygiene products?
- Which factors influence consumers in the purchase decision process when choosing sustainable and circular cosmetics and hygiene packaging designs?

2 Theoretical Background

Design for Sustainability (DfS) is the design discipline's response to the sustainability issues of our time. While earlier works primarily focused on reducing environmental impacts by redesigning individual features of products, more recent approaches have shifted to product-service systems, spatio-social, and socio-technical systems (Ceschin and Gaziulusoy 2016). Product-focused concepts like eco-design were the first to address the whole life cycle of products, i.e., from resource extraction, manufacturing, and use to the product's end-of-life (Ceschin and Gaziulusoy 2016). However, past empirical studies provide evidence that they have not achieved the decoupling of economic growth and natural resource use (e.g., United Nations Environment Programme (UNEP) 2011). That UNEP report

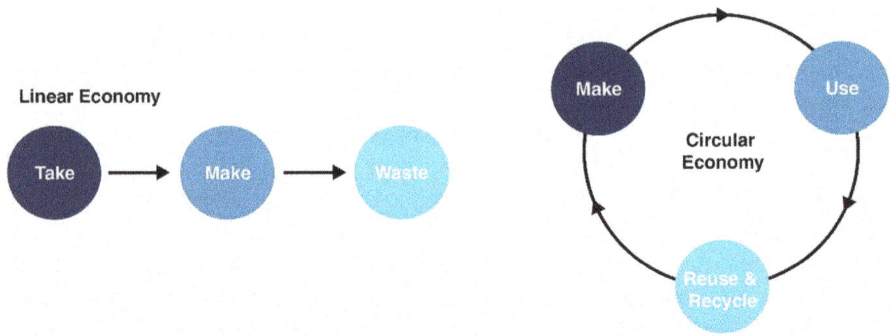

Fig. 1 Linear vs. circular economy based on UBA (2020)

showed that efforts to increase recycling quotas, energy efficiency levels or other strategies of eco-design have only provided incremental improvements often compensated by rebound effects and exaggerating growth of conventional markets. While these eco-design strategies remain valid, more powerful strategies must be implemented. As mentioned in Sect. 1, a transitional sustainability change may be the potential of CE (Reike et al. 2018).

The **circular economy (CE)** can be understood as a counter-concept to the linear economy ("take-make-waste") that has prevailed since the beginning of industrialization. Figure 1 illustrates the contrasting differences between both approaches (German Environment Agency (UBA) 2020). The CE approach aims to foster reuse within closed material loops rather than extracting resources as the basis for economic growth, thus reducing supply risks (Andrews 2015). Any system based on consumption rather than regenerative use of resources would otherwise cause significant losses along the value chain (Ellen MacArthur Foundation 2013).

CE has recently received a lot of attention but is interpreted differently (Circle Economy 2023; Corona et al. 2019; Kirchherr et al. 2017). A comprehensive definition of a **sustainable CE**, based on an analysis of 114 CE definitions, is provided by the latter author's team: it is

> "an economic system that replaces the 'end-of-life' concept with reducing, alternatively reusing, recycling, and recovering materials in production/distribution and consumption processes. It operates at the micro level (products, companies, consumers), meso level (eco-industrial parks), and macro level (city, region, nation, and beyond), with the aim to accomplish sustainable development, thus simultaneously creating environmental quality, economic prosperity and social equity, to the benefit of current and future generations. It is enabled by novel business models and responsible consumers." (Kirchherr et al. 2017, p. 229)

Many CE definitions often ignore sustainability, especially social sustainability, and do not directly address planetary boundaries (Melles et al. 2022a, b). We, therefore, distinguish between a (standard) CE and a sustainable CE. In order to determine the sustainability of a CE concept, one should examine not only its circularity, but assess its sustainability level separately (Hatzfeld et al. 2022).

Within dedicated literature, related **circular strategies** start with 'R' and encompass up to 10 'Rs' (Kirchherr et al. 2017; Reike et al. 2018). The long-established '3R' strategy set from waste management, e.g., relies on Reduce, Reuse, and Recycle. A more recent and comprehensive model complements the '3R' with Repair, Refurbish, Remanufacture, Repurpose, Refuse, and Reduce by Design (UNEP 2022, see also chapter 7 by Melles and Velenturf). However, according to Reike et al. (2018), the simpler '3R' typology was most frequently used in the past CE literature. Far-reaching CE strategies, e.g., Refuse or Rethink have gained little attention so far. In this analysis, the '4R' strategies, including Refuse, Reduce, Reuse, and Recycle, have been a guide to categorize the findings below.

In this paper, a **circular product design** is understood as a combination of different DfS approaches aiming to design products that are aligned with a sustainable CE. This relationship between DfS and CE is well described by Ceschin and Gaziulusoy (2016). A circular product design should make products easy to repair, reuse, disassemble, remanufacture, and/or recycle (Willskytt 2021). As an extended concept within circular product design, product-service systems comprise a product and a service in their basic elements. Companies shift their attention from developing and selling purely physical products to handling a system of products and services (Lofthouse and Bhamra 2006). We assume these conceptual considerations can also be transferred to a circular packaging design.

The success of circular design solutions depends on a number of boundaries, such as policies, economic constraints, and not least on **customers' acceptance**. Customers' purchase decisions are no longer (and never were) based solely on price and quality. Instead, the buying decision also depends on a product's ability to reinforce personal attributes (Irani and Frankel 2020). Irani and Frankel (2020) showed by neuromarketing methodology that emotions are crucial influencing factors. This is supported by Kahraman and Kazancoglu (2019) who argue that choosing more sustainable (greener) products can be explained by consumers' preference to buy products that are consistent with their values. Consumers with high levels of environmental awareness and trust in sustainable products are more likely to purchase products that convey the same values.

The **functions of packaging in general** are illustrated in Fig. 2. The product yields a value and usually has a larger footprint than the packaging. The product is protected by the packaging against conditions such as physical loads, biological contamination, chemical contamination, and last but not least, exposure to ultraviolet radiation (Paine and Paine 1992). The packaging keeps the value of the product along the life cycle and helps to inform and convince customers about and after its purchase. This is usually achieved by primary packaging (e.g., plastic tube or box, may contain a number of products), that is often complemented by secondary packaging (e.g., cardboard box; may contain numbers of primary packaging), and tertiary packaging to protect larger numbers of the products and their primary/secondary packaging during transport to the point of sale (Hanlon et al. 1998; Muntean et al. 2019).

All three packaging layers have an environmental footprint. Materials, additives, and energy are necessary to produce the packaging. At the end-of-life, materials have to be sorted, transported, and processed with further energy and resource consumption. Parts of the materials may be used as post-consumer recycled material for new packaging or other products. Usually, the larger part will be burnt to derive energy or heat from it, euphemistically referred to as thermal recycling. Even though the materials are taken from the cycle, the energy use can also have a positive impact as quantified in LCA. Packaging parts that could not be material or thermally recovered are usually landfilled. The latter two options are to be avoided in the sustainable circular economy. The packaging has the described footprints, but also protects the product and its value against the impacting conditions. As

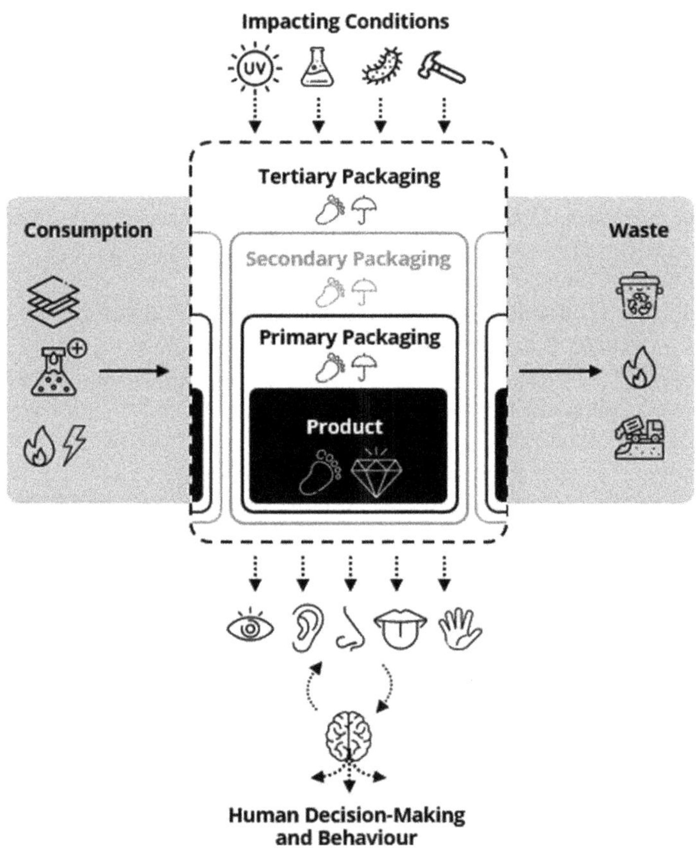

Fig. 2 Core functions of product packaging

Wikström et al. (2019) point out for the food domain, the "total environmental burden of product/package" (p. 534) needs to be assessed considering the packaging functions also as a means to reduce (product) waste.

The design and condition of packaging and products are perceived by customers through sensory channels. Customers holistically experience the offerings. This experience is characterized by the perception of (anticipated) instrumental qualities as well as non-instrumental qualities (e.g., convenience or 'joy of use')—always in constant interplay with emotions (Thüring and Mahlke 2007). This process is unconscious and influenced by human factors such as attitudes, needs, and the individual aims of each customer as well as the specific context. Thus, human judgments play a role in decision-making and behavior (Wölfel and Krzywinski 2019).

3 Methodology

In order to identify relevant academic publications, a systematic literature review was conducted based on Fink's (2019) methodology. The literature was extracted from the databases Scopus, Wiley, ScienceDirect, and EBSCOhost based on their thematic fit with the research topic. Search terms were selected from related topics of packaging design (e.g., 'packag*'), circular design (e.g., 'sustainab*'), and cosmetics (e.g., 'personal care') and applied to the title, abstract, and keywords (see Table 1). A total of 17 relevant publications were used for the literature review.

Practical search criteria were then selected and applied: English-language literature was included without publication date restrictions. Only types of double-blind peer-reviewed academic literature were considered, too. In a further step, the potential publications were subject to a methodological screening after removing duplicates based on the screening title, abstract, and full text. Some publications were removed due to lack of availability, inappropriate content, or low quality. That final screening step and forward/backward search techniques resulted in 17 relevant publications. The literature sample covers a period of publication year

Table 1 Used search terms within the systematic literature review

Packaging design	**AND**	**Circular design**	**AND**	**Cosmetics**
Product packag*		circular		cosmetic*
OR packag* design		**OR** sustainab*		**OR** "make up" OR makeup
		OR green OR eco*		**OR** "beauty product"
		OR recycl* OR reus*		**OR** "personal care" OR personalcare
		OR organic* OR bio*		**OR** "oral care" OR oralcare

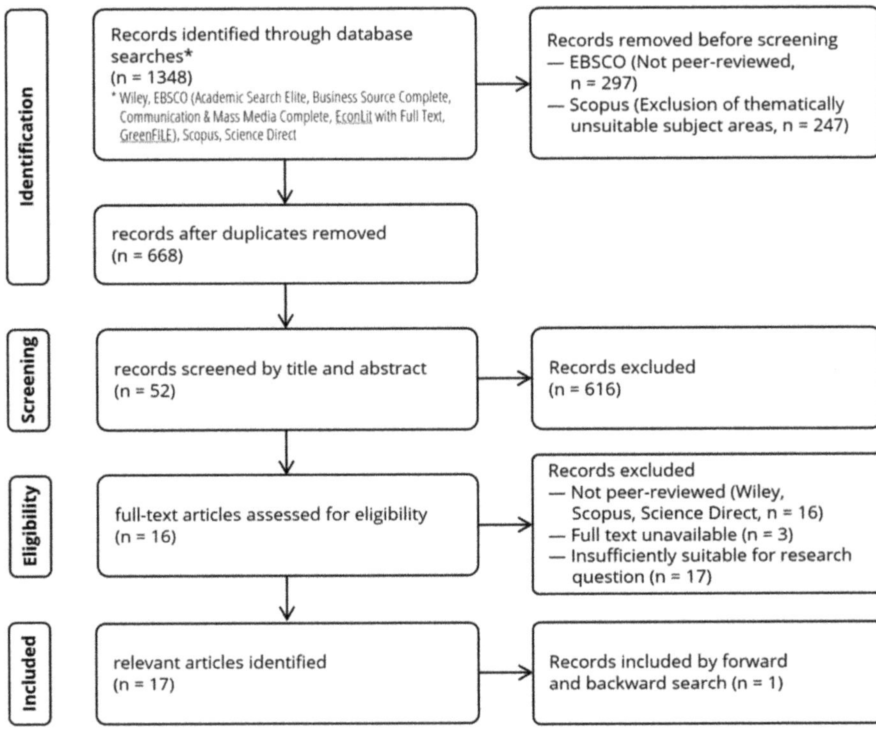

Fig. 3 Search scheme of the systematic literature review based on Page et al. (2021)

from 2006 to 2022 with the majority (70%) being published in the last six years. The search scheme is shown in Fig. 3.

The collected literature was analyzed by qualitative content analysis using the software MAXQDA according to Kuckartz and Rädiker (2019), which enables a systematic, replicable, and rule-guided analysis of text material. For the basic form of content-structuring qualitative content analysis, thematic categories were developed from research questions. The material was randomly coded according to that category system. Subsequently, text passages of the same categories were compiled and subcategories were determined based on the text material. For the categorization of packaging alternatives, we followed the 'R' strategies as laid out in Sect. 2. In the course of the iterative inductive coding process, the '3R' strategy set (i.e., Reduce, Reuse, and Recycle) was supplemented by the Refuse strategy.

4 Results

The results from the systematic literature review cover findings on general functions of packaging, conventional packaging concepts, and its circular, possibly more sustainable alternatives. We identify aspects of functions, design (features), materials, environmental impacts, disposal, and risks, among others, directly related to the packaging design for CE solutions. We then discuss potential sustainability criteria and what factors influence consumers' purchase decisions.

4.1 Core Functions of Packaging

The findings of the review correspond to the general description of core packaging functions as described in Sect. 2 and illustrated in Fig. 2. Packaging is a **by-product** and is not bought for its own sake (Lofthouse et al. 2017). It protects the product, prevents losses, and ensures quality or facilitates logistical processes (Aguiar et al. 2022; Jaccarini and Refalo 2017; Sahota 2013). This feature is worth highlighting in relation to the overall sustainability assessment discussed in this review. Against the backdrop that the environmental footprint of goods usually exceeds that of the packaging multiple times, an excessively high proportion of broken and unsaleable goods can thus lead to worse quantified environmental impacts using LCA in comparison to conventionally but well-packaged products (Pauer et al. 2019). Only a few authors explicitly mention the protection of the shielding against microorganisms or product tampering (Rosette et al. 2012).

In addition, customers can be informed about the correct use and function of the product with the help of information on the packaging (Aguiar et al. 2022; Jaccarini and Refalo 2017). In cosmetics, the packaging is also used for further communication (Aguiar et al. 2022; Lofthouse et al. 2017; Sahota 2013). It can help improve the product's differentiation and image and create marketing advantages (Rosette et al. 2012; Yablonski and Mancuso 2011). Graphic elements and information can be used to illustrate important features, which can be very extensive (Lofthouse et al. 2017). Information on improved use and disposal on the label significantly impacts purchase and actual use and can be provided as graphic guidance (Aguiar et al. 2022; Ren et al. 2022). Information on desired user behavior can reduce the loss of content (Willskytt 2021). At the same time, the packaging is an important task in communicating brand identity. Similar to mood and attention, colors influence differentiation within the product family and from the competition. In cosmetics, for instance, colors can follow the content so that the feeling of freshness and cleanliness is achieved (Malea et al. 2020).

4.2 Current Challenges of Conventional Packaging

There is a growing recognition that packaging designs that are not environmentally sustainable are a thing of the past (Yablonski and Mancuso 2011). However, most cosmetics currently on the market are still packaged in single-use plastic containers (Gatt and Refalo 2022; Lofthouse et al. 2017; Wakefield-Rann 2017)—a problem not only seen for hygienic and cosmetic products (Zhu et al. 2022).

Different cosmetic packagings have their most significant **environmental impact** at different product life cycle stages. Studies with LCA for cosmetics packaging provide quantitative results: Ren et al. (2022) conducted LCA for glass and plastic PET cosmetic bottles in China. They found that the end-of-life (EoL) has the highest environmental impact at over 50%, with landfilling of waste being the most severe, followed by the manufacturing, transport, and use stages. The LCA of Jaccarini and Refalo (2017) showed that in the case of a transportable cosmetic box with powder and mirror, the raw material extraction phase is the largest contributor to all life cycle stages when only life cycle energy consumption is considered. In the analysis by Civancik-Uslu et al. (2019), raw materials also accounted for half of an ordinary plastic tube's total life cycle impacts. However, this LCA was carried out excluding any modeling of EoL. The rationale was to convey the environmental profile of the tubes to the manufacturers. Waste disposal would vary depending on the specific markets where the tubes are sold. So, even the most environmentally friendly packaging is no good if there is no system to collect and recycle it at EoL (Sahota 2013). The difference between different countries in terms of their collection systems can be seen in Fig. 4.

Aguiar et al. (2022) direct the perspective additionally to the aspect of water consumption, whereby this is mainly incurred in the cleaning of production facilities, materials, and packaging lines, as well as for compliance with hygiene standards. The main problem with plastics is the impact of marine pollution on the environment, which has not yet been studied using LCA methods (Civancik-Uslu et al. 2019).

For many materials in the cosmetics sector, conventional disposal is often the only EoL solution considered, as there are currently no sufficiently proven, efficient, and cost-effective **technologies for recycling** materials such as laminates

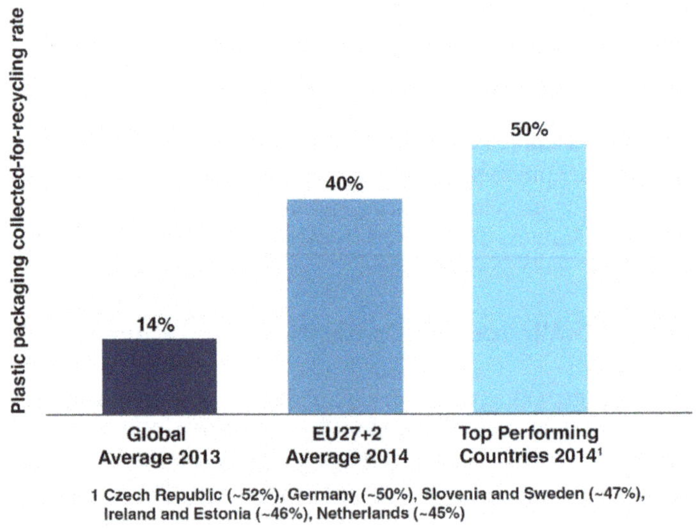

Fig. 4 Plastic packaging collected-for-recycling rates in different regions based on World Economic Forum et al. (2016, p. 49)

(Malea et al. 2020). While recycling aims to minimize a product's impact, it could problematically encourage the use of disposable products by giving the appearance that recyclable products cause less harm (Wakefield-Rann 2017). Some packages, like cosmetic tubes, are complicated to recycle as they still contain cosmetics in the waste and end up in the mixed plastic fraction at the packaging waste sorting plant due to their small size. Therefore, they are landfilled or incinerated in many countries (Civancik-Uslu et al. 2019).

As a measure against the loss of resources and the emissions caused by production and disposal, the Packaging Directive 94/62/EC was adopted in the EU in 2021 (Aguiar et al. 2022; Directive 94/62/EC 1994). This stipulates that packaging should be produced with minimum volume and weight, safety, hygiene, and consumer acceptance. Companies' packaging policy should be implemented sustainably and responsibly and build on the '3R' pillars. Biodegradability and responsible sourcing of materials should also be considered (Aguiar et al. 2022).

4.3 Current Challenges for Sustainable Packaging

Packaging design must ensure the core functions of packaging (see Sect. 4.1). These include the protection and the communicative means of primary, secondary, and tertiary packaging, which must be achieved under specific material, process, market, and other constraints. The packaging design had been a complex challenge already before the commencing sustainable transition. For example, secondary packaging is often not deployed. In such cases, primary and tertiary packaging must provide all packaging functions. The shift to sustainable packaging brings even more design challenges to be accomplished.

The design phase is crucial for each life cycle phase of cosmetics, as it defines the whole system. Design has the potential to influence purchase and consumption and can promote use behavior, reuse, and recycling (Ren et al. 2022). According to Lofthouse and Bhamra (2006), the **goal of an alternative design** must be to add value to a product by making it convenient for customers. Ease of use should be provided without compromise (Rosette et al. 2012). According to these authors, the packaging system should also be considered in terms of product characteristics, sterilization methods (if applicable), sealing, labeling, secondary packaging, handling, shipping, environment, storage, government regulations, and end use.

Design methods that improve the product's environmental performance should be used to reach these goals. Checklists/guidelines, diagramming tools, CAD programs, and LCA can be used integratively for this purpose (Willskytt 2021). LCA was used in many of the studies reviewed, and the importance of using LCA in combination with eco-design strategies to improve the environmental profile was also addressed. This can help identify key life cycle stages where eco-design strategies are most efficient (Civancik-Uslu et al. 2019; Jaccarini and Refalo 2017).

A good overview of **eco-design guidelines** is provided by Willskytt (2021). The review of the literature on eco-design and the CE examined the applicability of design guidelines for resource-efficient products to consumer products. It was identified that most design guidelines exist for the packaging, but only a few for cosmetic products. However, several of the identified guidelines are transferable to cosmetic products. In particular, these aspects should be taken into consideration: Packaging sizes should fit the user; dispensing quantities should be adapted to user behavior; the dispensing capability of the packaging should be optimized so that the entire contents can be dispensed; information can be provided to help reduce energy and water consumption. In addition, it can be stated that design considerations can be transferred from products with similar levels of required packaging protection and shelf life. This would allow the application of considerations for canned food products to cosmetic products. Furthermore, for products used by similar actors, design considerations for product handling can often be transferred (Willskytt 2021).

Structural designs invite people to touch the product (Sahota 2013). However, superfluous, non-functional components and voids should be eliminated to save material. Structural changes such as reducing, strengthening, or weakening components can be used for this purpose (Willskytt 2021).

Rising oil prices already put pressure on companies to **use alternative materials** for plastic packaging (Sahota 2013). In addition, regulations further promote recycled, reusable or recyclable plastics in a cost-effective way (Aguiar et al. 2022). Sustainable and resource-efficient use of materials can be accomplished by avoiding scarce, critical, non-biodegradable materials and substituting them with less environmentally burdensome materials (Willskytt 2021). Here, however, it is important to point out that the sustainability of biodegradable materials is highly controversial (Filiciotto and Rothenberg 2021; UBA 2017). The German UBA advises against their use in the packaging sector: Single-use products are criticized as short-lived and generate unnecessary waste, regardless of whether they are biodegradable. Disposing of biodegradable single-use items in the environment has similar negative consequences as compared to conventional plastics, as these materials degrade only very slowly in fields and meadows. UBA assumes that indications of biodegradability may even encourage people to careless and inconsiderate waste handling (UBA 2017). Materials that are difficult to separate, such as laminates and composites, should, if possible, be avoided. Instead, mono materials could be used in the product to increase recyclability (Ren et al. 2022; Willskytt 2021).

The awareness of many companies to switch to **natural, organic materials** also poses another risk. It may provide the opportunity for better interaction between packaging and product, as a number of antimicrobial agents are naturally present in organic materials. But it may also provide for possible inactivation of preservatives, contributing to faster degradation of the preservative system. Environmentally friendly substances in packaging and formulation may also require the addition of preservatives, as the previously high pH and hostile raw materials were more resistant to microbes and now make the product more vulnerable (Yablonski and Mancuso 2011).

There are significantly fewer design guidelines for food and cosmetics packaging than those for medical devices (Willskytt 2021). But what **learnings in other**

industries could inspire packaging design considerations related to the cosmetics industry? Similar to food, there are also expiry dates for cosmetic products. However, that is insufficiently communicated on the packaging as European regulation requires to inform consumers only if the shelf life is less than 30 months (Willskytt 2021). That information is essential as prolonged use of the product is linked with a decrease in preservatives and an increase in microbial **contamination**. Two main risk factors prevail: First, a misuse of cosmetics for another purpose than intended. Second, product accessibility and exposure, i.e., the contact of an organism with chemical, biological or physical agents (Rosette et al. 2012; Yablonski and Mancuso 2011).

The publications of Rosette et al. (2012) and Yablonski and Mancuso (2011) summarize findings for design aspects to address these issues. For example, smaller packages are safer, i.e., reduce the likelihood of contamination, because of a shorter use period and lower frequency of interaction. Moreover, packaging with larger dispensing openings such as jars, bottles, and cosmetic cans tend to be more vulnerable than aerosol cans, airless pumps, or sealed systems. Obviously, safety considerations contradict the previously discussed ideas for reducing the amount of packaging. Generally, the applicator and delivery mechanism are considered the most critical components in terms of product exposure. For applicators, sponge applicators and mascara brushes are at high risk, whereas dry powder brushes, swabs, puffs, and disposable sponges are the most suitable. However, there is no single strategy to minimize applicator risk. An ideal opening is one with easy access and low exposure to the environment. A consumer product evaluation study showed that contamination heavily depends on user scenarios: a flip-cap closure for shampoos was found to be 0% contaminated after use, while the same closure for a hand cream was 39% contaminated because consumers used the product differently. (Rosette et al. 2012; Yablonski and Mancuso 2011).

Similar detailed information on contamination risks were not presented in the reviewed literature. Aguiar et al. (2022) indicate that cosmetics with water are susceptible to microbes. Amberg and Fogarassy (2019) support these findings, adding cosmetics with organic mixtures as vulnerable. Willskytt (2021) only refers to carefully disposing of contaminated waste with biological materials, borrowing from medical products. Lofthouse and Bhamra (2006) suggest adopting mechanisms from a medical design to ensure that packaging communicates the harmful consequences of using the product in concentrated form or avoiding it.

4.4 Design Considerations for Circular Packaging and Its Role in Sustainability

Sustainable and new practices are becoming increasingly important among product developers (Gatt and Refalo 2022). New approaches to design, such as product-service systems, could be applied. These aim to achieve integrated functional solutions to meet customer needs. For example, detergent manufacturers go door-to-door with delivery vans and supply their customers by each taking the amount they need in their own container (Lofthouse and Bhamra 2006). Despite that, different approaches based on the 'R' strategies have been identified in the literature as presented in the following. In consent with Zhu et al. (2022), the CE strategies that are implemented in 'sustainable cosmetics packaging' remain close to the core '3R' strategy of Reduce, Reuse, and Recycle. Hence, they stick with the rather conventional approach of eco-design.

The more sufficiency-oriented strategy of Refuse has been discussed as applicable to sustainable cosmetics packaging. Other strategies such as Rethink or Refurbish have not explicitly been mentioned. However, Rethinking as such might be hardly apparent in current forms of academic literature. Refurbishment in packaging design might be hard to differentiate from Reuse or Recycling. Even the long-established refurbishment of dairy goods glass bottles is commonly recognized under the umbrella of Recycling. However, such kind of strategy has not been identified in the review for use in cosmetics packaging. An overview of the different 'R' strategies, both identified in the selected literature and not yet discussed is illustrated in Fig. 5. Within the illustration, the Recover strategy is assigned to Recycle for simplification.

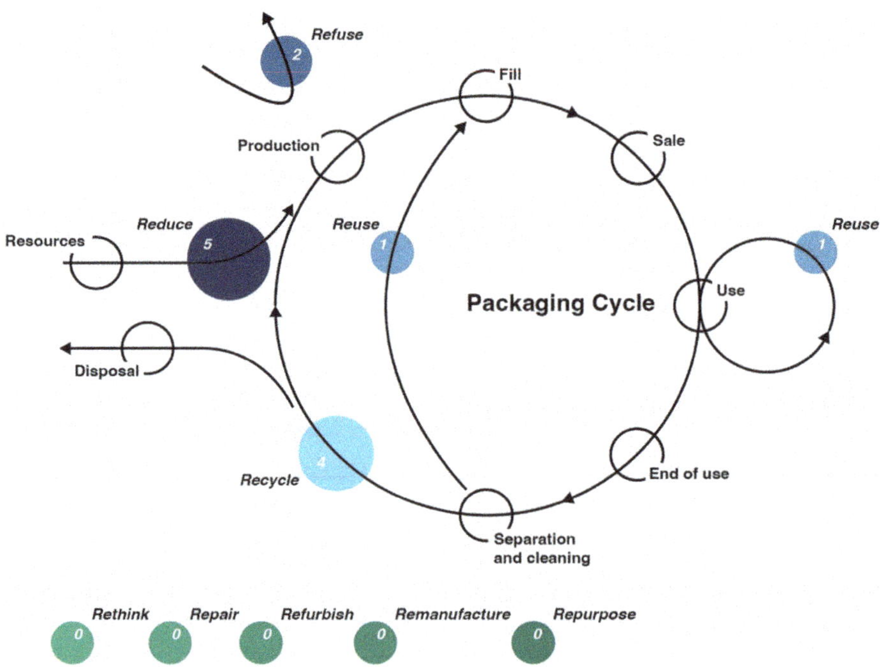

Fig. 5 'R' strategies for sustainable cosmetics packaging as identified in the literature review (blue) and other potential strategies not yet discussed in the sustainable cosmetic packaging literature (green)

4.4.1 Refuse

The strategy that is most compatible with the goals of a sustainable CE is the avoidance of packaging. In this case, the product is designed in such a way that there is no need for packaging at all. For instance, anhydrous formulas offer the optimal conditions for this (Aguiar et al. 2022). Although highly concentrated, water-free products appear more expensive to customers, they last longer and replace three to four equivalent non-water-free products, reducing purchase frequency and increasing savings (Aguiar et al. 2022). A prime example of this sales strategy is the company Lush, which claims to sell 65% of its products "naked" and also uses this strategy in marketing campaigns (Aguiar et al. 2022; Gatt and Refalo 2022; Sahota 2013; Wakefield-Rann 2017). This has enabled them to save more than 450,000 L of water per year compared to liquid shampoos. However, it should be noted that according to EU Regulation No. 1223/2009, non-pre-packaged products need to be accompanied by product information (Aguiar et al. 2022).

At the same time, it is important to emphasize that packaging, as described above, has a protective function. Accordingly, neglecting effective packaging can lead to a greater amount of broken goods. Since the ecological footprint of the goods usually exceeds that of the packaging by a multiple, an excessively high proportion of unsaleable goods can thus lead to worse LCA results than the same well-packaged item (Pauer et al. 2019). Dissolvable packaging is another concept, that can be categorized as a refuse strategy (Willskytt 2021).

4.4.2 Reduce

The reduction strategy primarily focuses on reducing the amount of material through downsizing, weight reduction, structural changes such as reinforcing, folding, splinting, framing, minimizing thickness, avoiding low-function components, and minimizing the number of separable components that could end up in waste (Willskytt 2021). An LCA shows that weight reduction can reduce the overall impact of a plastic tube by 10% on average (Civancik-Uslu et al. 2019). However, the product should be tested for stability, which can be affected by dematerialization (Gatt and Refalo 2022).

A differentiated concept is the use of larger packaging. This concept should lead to simplification and reduction of the relationship between product and packaging (Aguiar et al. 2022). However, this concept should be viewed critically. Redesign approaches should be encouraged but do not have a radical impact on the LCA. While the weight of packaging per unit has decreased, demographic changes, the size of families, and the demand for more convenience have led to an increase in the amount of packaging used (Lofthouse and Bhamra 2006). The use of larger packaging can also be seen as critical in terms of possible contamination, as discussed earlier.

4.4.3 Reuse

Strategies aimed at reusing products are not as widespread as switching to sustainable materials (Sahota 2013). The packaging guidelines refer to reuse as a process whereby a package is designed to go through a minimum number of cycles during its life cycle or can be re-filled, with or without the help of auxiliary products on the market that enable refilling. Such packaging becomes packaging waste when it is no longer reusable (Aguiar et al. 2022; Gatt and Refalo 2022).

If used as expected, returnable packaging systems can provide greater environmental benefits and support the transition to a sustainable CE, which advocates the development of circular material flows (Lofthouse et al. 2017). Gatt and Refalo (2022) demonstrate that the positive impact of reusability can exceed that of dematerialization by 171%, even if the plastic waste of reusable packaging is not recyclable. Moreover, applying recyclability to a product that is already reusable does not significantly reduce the environmental impact. However, reuse should not necessarily be used as the only strategy. The study by Gatt and Refalo (2022) additionally showed that a cosmetic can with a replaceable aluminum tray containing the product that has to be repurchased is less sustainable than repeatedly buying the same can without the aluminum tray. In this case, reusability does not compensate for the negative impact of the aluminum tray. Furthermore, there is an overall risk with all reusable products that consumers will dispose of this type of product (Wakefield-Rann 2017), and therefore, even lead to higher resource and energy consumption, as this type of packaging is likely to be heavier as it needs to last longer (Lofthouse et al. 2017).

In addition to the reusability challenges already mentioned, a product should be adapted to consumer needs, as in the past the use of reusable packaging has not been successful in some cases (Lofthouse and Bhamra 2006; Yablonski and Mancuso 2011). A high level of inconvenience and low incentives were seen as the cause. This problem can be addressed by the concept of emotionally durable design, which aims to build an emotional bond between the product and the user so that he is less inclined to replace it (Willskytt 2021). Attractive packaging that is valuable in the eyes of the consumer will encourage them to refill and keep it rather than throw it away (Lofthouse and Bhamra 2006). Limited acceptance can also be caused by negative associations such as poor quality and inconvenience (Lofthouse et al. 2017).

Packaging design needs to consider a number of technical issues related to durability, communication, refill mechanism, safety, and cleaning. An ideal system should be simple, intuitive, and inclusive in refilling (Lofthouse et al. 2017). In addition to usefulness, the aspect of beauty and pampering associated with the appeal of cosmetics must be promoted (Yablonski and Mancuso 2011). To ensure the efficiency of the system, it is also necessary to determine the frequency of use, because the system is only successful if the customer also returns the packaging (Lofthouse and Bhamra 2006).

New pack types can be unfamiliar to the consumer, so the designer's main task is to develop an innovative overall concept. For the consumer, it must be clear from the product that it is a refillable system to avoid customer confusion and an increase due to unintentional waste. Reusability must be clear both at the point of sale and when the product is used (Lofthouse et al. 2017). Safety aspects, in particular, must be quickly apparent to the customer, as products should not be used in concentrated form (Lofthouse and Bhamra 2006). Dispenser units and primary packaging need to be appropriately durable, as customers need to buy the unit as well as the product (Lofthouse et al. 2017). The design of the system is therefore crucial, as otherwise, it could be more financially attractive to dispose of the product and buy a new one (Lofthouse and Bhamra 2006).

The packaging should allow the extraction of the entire content (Malea et al. 2020). Easy cleaning should be possible for reuse (Ren et al. 2022). It must be clearly communicated to consumers how the system should be treated and which components should remain in the system and how. This is also crucial for material selection, as glass and PET are popular materials in much of Europe that could potentially be recycled (Lofthouse and Bhamra 2006). In the best case, maintenance should be minimal or not required at all (Lofthouse et al. 2017), which can be accomplished by design, for example, if the packaging has a smooth surface (Willskytt 2021).

Since large openings are more vulnerable from a microbial safety perspective, a refillable package should be designed to restrict access and protect the product and the consumer from overexposure and cross-contamination by minimizing direct access to the nozzle. Automatic removal or sealing of the pouch including a new dispensing device on the refill packaging would also be conceivable (Rosette et al. 2012; Yablonski and Mancuso 2011).

4.4.4 Recycling

Using recycled materials for a single-use product can lead to significant environmental savings of more than 90% by using existing materials, reducing the total amount of waste sent to landfills and reducing energy resources (Gatt and Refalo 2022). Despite these benefits, recycling is still not the most popular EoL option due to the limited circularity, as mentioned above, and the potential contamination at the end-of-use phase that makes it difficult to recycle a product (Gatt and Refalo 2022; Potting et al. 2017). Therefore, the strategy to design with recycled materials has some limitations.

Furthermore, studies show that there is a significant lack of recycling. A 2018 EU study found that only 41.5% of plastic waste generated in Europe is material recycled, while the rest ends up in landfills or is used exclusively for energy recovery (Gatt and Refalo 2022). Development should take into account the convenience of customers. If returning-for-recycling procedures are not easy, consumers will not return packaging properly (Aguiar et al. 2022).

A recycling-friendly design should consider the dismantlability of the recycling process. This can be realized by separable components and fewer materials. In addition, barrier-free instructions demonstrate proper recycling (Ren et al. 2022). Technical recyclability also depends on the choice of materials and should be compatible with recycling techniques (Willskytt 2021).

If the packaging is destined for incineration or landfill, some authors argue that it should at least be designed to be truly incinerable (avoiding ceramics or the like) and avoid hazardous additives, such as those found in polyvinylchloride (PVC) (Willskytt 2021).

4.4.5 Overview

Table 2 provides an overview of the various intervention strategies and related design considerations mentioned above, that can be assigned to R strategies.

4.5 Acceptance of Alternative Packaging Designs at the Purchase Decision

Decisions depend on differently weighted factors of each consumer (Amberg and Fogarassy 2019). Among the factors that can have an impact on the customer's decision-making are environmental awareness, packaging design, greenwashing, brand identity, and price.

In relation to product acceptance, packaging is a very influential tool (Mosleh-pour et al. 2021). Recent research often suggests that its communication and interaction must lead to highlighting the environmental benefits of the product (Ren et al. 2022). Information on the label about the sustainability of the product and suggestions on how to use it can influence purchases as customers have been shown to have limited knowledge of sustainable products (Aguiar et al. 2022). Ren et al. (2022) found that plastics are underestimated and glass and biodegradable plastics are overestimated by average consumers.

As convenience and effort play a crucial role, customers will opt for small, lightweight packs. Refill packs, which are more suitable for transport, take up less space and are easy to reuse (Lofthouse et al. 2017) will be preferred (Yablonski and Mancuso 2011). Ideally, handling should be a good experience for the customer (Lofthouse et al. 2017). Structural packaging and graphic elements can invite you to touch, turn, and feel the product. Packaging design is expected to convey the product's personality through its shape and form and to arouse emotions in the process (Sahota 2013). The Lush company tries to create a personal bond between customer and product by putting stickers of the product producer on the products (Wakefield-Rann 2017). They have also increased the sensory value of their products by allowing the customer to experience the product without packaging. However, sealed, opaque packaging can appear more hygienic to the customer (Wakefield-Rann 2017). Customers also have the desire to test the fragrances in the shops. The unpackaged products would solve this problem, whereas this is impossible with sealed or dissolvable refills (Lofthouse et al. 2017).

5 Conclusions

Current production and consumption patterns must change substantially to address the challenges of highly stressed planetary boundaries (Rockström et al. 2021). SDG #12 aims to prevent waste and encourage sustainability and CE. As presented in previous sections, only selected CE strategies with an integrated sustainability assessment could be key to that fundamental transformation. Many authors point

Table 2 Design strategies and considerations in relation to the '4R' strategies

Refuse	Reduce	Reuse	Recycle
Design strategies			
Avoid packaging	Material reduction through downsizing	Returnable packaging for refilling	Design with recycled materials
Dissolvable packaging	Material reduction through weight reduction	Design for refilling by the customer	Design for recycling
	Material reduction through structural changes		Design for separate return (and recycle)
	Avoiding low-function components		Design for incineration
	Minimizing the number of separable components		
	Larger packaging		
Design aspects/considerations			
Protective function	Stability	Possible disposal instead of reuse	Potential contamination
	Demand for convenience	Frequency of reuse	Dismantlability
	Contamination	Demand for convenience	Separable components
		Attractive packaging	Fewer Materials
		Durability	Disposal information
		Cleanability	Use of technically recyclable materials
		Simple refillability	Incinerability
		Refillability of the product must be understood	Avoiding hazardous additives
		Safety	
		Financial attractiveness of reusing the packaging	
		Drainability	
		Simple and clearly communicated maintenance	
		Contamination	

out that the social goals of sustainability are often neglected (Corona et al. 2019; Hatzfeld et al. 2022; Reike et al. 2018).

In particular, past reviews on CE show that recycling strategies are not sufficient, but other CE approaches to design to meet redesign, refuse, and reuse strategy paths may do so. Moreover, discourses around the CE often focus on ideas around recycling, which moves only a little away from an ‘end-of-pipe’ thinking. Accordingly, more far-reaching demands are being made for a move toward a comprehensive sustainable transformation of the economy and society. The demand for strong sustainability must shape consumption patterns and business models, and there must be a rethink in the area of political intervention. In particular, sufficiency strategies and product-service systems must also be more central (Melles et al. 2022a, b; Velenturf and Purnell 2021; Wilts and von Gries 2015). Other management control instruments like carbon offsetting do not add to that transformation, too, as a recent study discussed (Bergero et al. 2023).

Our review aimed to summarize and discuss the current state of art packaging (design) for the consumer-oriented cosmetics and hygiene products industries. Our first research question referred to identifying the functional role of packaging for cosmetics and hygiene products. The aim was to gain a basic understanding of the challenge of meeting these requirements on the one hand and the criteria for sustainable CE packaging design on the other (see Sect. 4.1 to 4.3). Second research question, it was found that a majority of the methods can be found in the area of the Reuse strategy. Crucial to success is the consideration of the whole (product-service) system, especially the type of reuse. Crucial is also the disposal of the products. As long as consumers do not know under which conditions the products have a sustainable benefit and the manufacturers do not offer a disposal system, there is a risk that the products will be disposed of in a conventional way (Sahota 2013). Safety is an important decision point for consumers who are skeptical about new products (Kahraman and Kazancoglu 2019).

However, most of the analyzed studies ignore the safety risk due to microbial contamination, as only two deal with specific design features to reduce a risk (Rosette et al. 2012; Yablonski and Mancuso 2011). Awareness needs to be raised here for further research, as criticism has been voiced, particularly in relation to recyclable packaging. The importance of user testing and user-centered development was identified as important in this regard, as well as serving to establish user acceptance. The strategy of avoiding packaging is less represented in the publications, which is due to the fact that most works look for alternatives rather than avoiding them. Moreover, this strategy may work for shampoo or shower gel, but becomes a challenge for transportable cosmetics and would require additional packaging.

The third research question was to identify factors that influence the purchase decision making process and how these can be implemented through design to enhance the positive impact of green sales. Based on the amount of theory found, there seems to be little research on decision-making for sustainable cosmetics in combination with design recommendations. Five factors were identified that can influence the purchase decision process to varying degrees, and subjective perceptual preferences should not be ignored. The investigated factors of design, brand identity, price, environmental awareness, and manipulation perception can provide intentions for how they can be used by the appropriate design.

The price factor in combination with environmental awareness does not provide any directly obvious derivations for the design. It depends on the environmental awareness of the consumer. Those who have a high environmental awareness will be willing to pay more and even demand it because it is a sign of quality for them. The rest of the consumers will demand a similar price as for conventional products. The manufacturers of refill packs partly rely on financial incentives through cheaper overall packaging including the refill pack to motivate all those with a lower willingness to pay to buy. At the same time, it could also provide an incentive for environmentally conscious customers to buy the packaging anew as a result. However, with the method of emotional, long-lasting design, the designer can ensure that customers do not exchange the refillable containers for new ones by offering the customer added value through reuse that they cannot buy, for example.

So what can we learn from past research on packaging in the cosmetics and hygiene products industries? Packaging design can increase the circularity of products and make products appear more attractive to customers. Concepts such as emotional design can offer customers added value and strengthen the bond with the products. Supporting information for sustainable use can be conveyed more easily through visualizations. Furthermore, product-service systems will gain importance in the future, under the aspect of system thinking, and will change the concepts of the cosmetics and hygiene industry in their development in the long term, as is already the case in other industries. The literature provides a number of approaches for the design of sustainable packaging, but often the statements on the design of cosmetic and hygiene products are rather less concrete. Accordingly, they rely on the transfer of design guidelines from other sectors such as the food or medical and pharmaceutical industries (Willskytt 2021).

Future studies should closely examine those 'R' strategies that are yet missing in past research. There is a lack of case studies illustrating their practical application. New research could collect consumer evaluations based on a larger number of LCA studies and prototype developments to assess purchasing potential. In addition, the effects of different points of sale could be investigated more closely, since online retailing does not allow for a sensory evaluation by the customer, and the influence of social media marketing and influencer marketing could also be added.

References

Aguiar JB, Martins AM, Almeida C, Ribeiro HM, Marto J (2022) Water sustainability: a waterless life cycle for cosmetic products. Sustain Prod Consum 32:35–51. https://doi.org/10.1016/j.spc.2022.04.008

Amberg N, Fogarassy C (2019) Green consumer behavior in the cosmetics market. Resources 8(3), Article 3. https://doi.org/10.3390/resources8030137

Andrews D (2015) The circular economy, design thinking and education for sustainability. Local Econ 30(3):305–315. https://doi.org/10.1177/0269094215578226
Bergero C, Gosnell G, Gielen D, Kang S, Bazilian M, Davis SJ (2023) Pathways to net-zero emissions from aviation. Nat Sustain, 1–11.https://doi.org/10.1038/s41893-022-01046-9
Ceschin F, Gaziulusoy I (2016) Evolution of design for sustainability: from product design to design for system innovations and transitions. Des Stud 47:118–163. https://doi.org/10.1016/j.destud.2016.09.002
Circle Economy (2023) The circularity gap report 2023: a circular economy to live within the safe limits of the planet. Circle Economy. https://www.circularity-gap.world/2023
Circular Economy Initiative Deutschland (2021) Circular economy roadmap for Germany. acatech/SYSTEMIQ. https://doi.org/10.48669/ceid_2021-3
Civancik-Uslu D, Puig R, Voigt S, Walter D, Fullana-i-Palmer P (2019) Improving the production chain with LCA and eco-design: application to cosmetic packaging. Resour Conserv Recycl 151:104475.https://doi.org/10.1016/j.resconrec.2019.104475
Corona B, Shen L, Reike D, Rosales Carreón J, Worrell E (2019) Towards sustainable development through the circular economy—a review and critical assessment on current circularity metrics. Resour Conserv Recycl 151:104498. https://doi.org/10.1016/j.resconrec.2019.104498
Dinh TCT, Lee Y (2021) "I want to be as trendy as influencers"—how "fear of missing out" leads to buying intention for products endorsed by social media influencers. J Res Interact Mark 16(3):346–364. https://doi.org/10.1108/JRIM-04-2021-0127
Elkington J (2018) 25 years ago i coined the phrase "triple bottom line". Here's why it's time to rethink it. Harv Bus Rev 25:2–5
Ellen MacArthur Foundation (2013) Towards the circular economy Vol. 1: an economic and business rationale for an accelerated transition. EMF. https://ellenmacarthurfoundation.org/towards-the-circular-economy-vol-1-an-economic-and-business-rationale-for-an
European Parliament and Council Directive 94/62/EC of 20 December 1994 on packaging and packaging waste (1994). http://data.europa.eu/eli/dir/1994/62/oj
Filiciotto L, Rothenberg G (2021) Biodegradable plastics: standards, policies, and impacts. Chemsuschem 14(1):56–72. https://doi.org/10.1002/cssc.202002044
Fink AG (2019) Conducting research literature reviews: from the internet to paper, 5th ed. SAGE Publications
Finkbeiner M, Schau EM, Lehmann A, Traverso M (2010) Towards life cycle sustainability assessment. Sustainability 2(10), Article 10. https://doi.org/10.3390/su2103309
Gatt IJ, Refalo P (2022) Reusability and recyclability of plastic cosmetic packaging: a life cycle assessment. Resour Conserv Recycl Adv 15:200098. https://doi.org/10.1016/j.rcradv.2022.200098
Geels FW (2011) The multi-level perspective on sustainability transitions: responses to seven criticisms. Environ Innov Soc Trans 1(1):24–40. https://doi.org/10.1016/j.eist.2011.02.002
Geissdoerfer M, Savaget P, Bocken NMP, Hultink EJ (2017) The circular economy—a new sustainability paradigm? J Clean Prod 143:757–768. https://doi.org/10.1016/j.jclepro.2016.12.048
German Environment Agency (UBA) (26 Oct 2017). Biobasierte und biologisch abbaubare Kunststoffe [only in German] [Text]. Umweltbundesamt. https://www.umweltbundesamt.de/biobasierte-biologisch-abbaubare-kunststoffe
German Environment Agency (UBA) (2020) 9 principles for a circular economy. UBA. https://www.umweltbundesamt.de/publikationen/9-principles-for-a-circular-economy
Hanlon JF, Kelsey RJ, Forcinio HE (1998) Handbook of package engineering, 3rd ed. Taylor & Francis/CRC Press
Hatzfeld T, Backes JG, Guenther E, Traverso M (2022) Modeling circularity as Functionality Over Use-Time to reflect on circularity indicator challenges and identify new indicators for the circular economy. J Clean Prod, 134797. https://doi.org/10.1016/j.jclepro.2022.134797
Irani R, Frankel L (2020) The role of design in consumer behaviour: how design can influence consumer decision making at a point of purchase. In: Di Bucchianico G, Shin CS, Shim S, Fukuda S, Montagna G, Carvalho C (eds) Advances in industrial design. Springer International Publishing, pp 316–322. https://doi.org/10.1007/978-3-030-51194-4_42

Jaccarini P, Refalo P (2017) Life cycle analysis of a portable cosmetic package. In: Challenges for technology innovation: an agenda for the future. CRC Press

Kahraman A, Kazancoglu I (2019) Understanding consumers' purchase intentions toward natural-claimed products: a qualitative research in personal care products. Bus Strateg Environ 28(6):1218–1233. https://doi.org/10.1002/bse.2312

Kirchherr J, Reike D, Hekkert M (2017) Conceptualizing the circular economy: an analysis of 114 definitions. Resour Conserv Recycl 127:221–232. https://doi.org/10.1016/j.resconrec.2017.09.005

Kuckartz U, Rädiker S (2019) Introduction: analyzing qualitative data with software. In: Kuckartz U, Rädiker S (eds) Analyzing qualitative data with MAXQDA: text, audio, and video. Springer International Publishing, pp 1–11. https://doi.org/10.1007/978-3-030-15671-8_1

Lofthouse V, Bhamra TA (2006) Refillable packaging systems: design considerations. In: Proceedings of international design conference-DESIGN 2006, 15–18 May. https://hdl.handle.net/2134/2188

Lofthouse V, Trimingham R, Bhamra T (2017) Reinventing refills: guidelines for design. Packag Technol Sci 30(12):809–818. https://doi.org/10.1002/pts.2337

Malea A, Tzotzis A, Manavis A, Kyratsis P (2020) Innovative and sustainable toothpaste packaging design. J Graph Eng Des 11(2):19–30. https://doi.org/10.24867/JGED-2020-2-019

Melles G, Wölfel C, Krzywinski J, Opeskin L (2022) Expert and diffuse design of a sustainable circular economy in two German circular roadmap projects. Soc Sci 11(9), Article 9. https://doi.org/10.3390/socsci11090408

Melles G, Wölfel C, Opeskin L (2022) Technology design for a sustainable circular economy: research and practice consequences. In: Leal Filho W, Azul AM, Doni F, Salvia AL (eds) Handbook of sustainability science in the future: policies, technologies and education by 2050. Springer International Publishing, pp 1–20. https://doi.org/10.1007/978-3-030-68074-9_141-1

Moslehpour M, Chaiyapruk P, Faez S, Wong W-K (2021) Generation Y's sustainable purchasing intention of green personal care products. Sustainability 13(23), Article 23. https://doi.org/10.3390/su132313385

Muntean AC, Juncan AM, Moisa DG, Vonica AL, Rus LL, Morgovan C, Gligor FG, Butuca A, Stanila A (2019) Primary packaging and stability evaluation of a serum used for the periorbital area of the sensitive eye. Mater Plast 56(2):360–365. https://doi.org/10.37358/MP.19.2.5186

Page MJ, McKenzie JE, Bossuyt PM, Boutron I, Hoffmann TC, Mulrow CD, Shamseer L, Tetzlaff JM, Akl EA, Brennan SE, Chou R, Glanville J, Grimshaw JM, Hróbjartsson A, Lalu MM, Li T, Loder EW, Mayo-Wilson E, McDonald S, Moher D et al (2021) The PRISMA 2020 statement: an updated guideline for reporting systematic reviews. BMJ 372, n71. https://doi.org/10.1136/bmj.n71

Paine FA, Paine HY (1992) A handbook of food packaging. Springer US.https://doi.org/10.1007/978-1-4615-2810-4

Pauer E, Wohner B, Heinrich V, Tacker M (2019) Assessing the environmental sustainability of food packaging: an extended life cycle assessment including packaging-related food losses and waste and circularity assessment. Sustainability 11(3), Article 3. https://doi.org/10.3390/su11030925

Plastics Europe, & EPRO (2022) Plastics—the facts 2021. Plastics Europe. https://plasticseurope.org/knowledge-hub/plastics-the-facts-2021/

Potting J, Hekkert M, Worrell E, Hanemaaijer A (2017) Circular economy: measuring innovation in the product chain (Policy Report No. 2544; PBL Publication). Netherlands Environmental Assessment Agency (PBL). www.pbl.nl/sites/default/files/cms/publicaties/pbl-2016-circular-economy-measuring-innovation-in-product-chains-2544.pdf

Reike D, Vermeulen WJV, Witjes S (2018) The circular economy: new or Refurbished as CE 3.0?—exploring controversies in the conceptualization of the circular economy through a focus on history and resource value retention options. Resour Conserv Recycl 135:246–264. https://doi.org/10.1016/j.resconrec.2017.08.027

Ren Z, Zhang D, Gao Z (2022) Sustainable design strategy of cosmetic packaging in china based on life cycle assessment. Sustainability 14(13):8155. https://doi.org/10.3390/su14138155

Rockström J, Gupta J, Lenton TM, Qin D, Lade SJ, Abrams JF, Jacobson L, Rocha JC, Zimm C, Bai X, Bala G, Bringezu S, Broadgate W, Bunn SE, DeClerck F, Ebi KL, Gong P, Gordon C, Kanie N, et al (2021) Identifying a Safe and Just Corridor for People and the Planet. Earth's Future 9(4):e2020EF001866. https://doi.org/10.1029/2020EF001866

Rodriguez-Anton JM, Rubio-Andrada L, Celemín-Pedroche MS, Alonso-Almeida MDM (2019) Analysis of the relations between circular economy and sustainable development goals. Int J Sust Dev World 26(8):708–720. https://doi.org/10.1080/13504509.2019.1666754

Rosette JL, Yablonski J, Mancuso S, Kale G (2012) Packaging, cosmetics and pharmaceuticals. In: Kirk-Othmer encyclopedia of chemical technology. Wiley, pp 1–14. https://doi.org/10.1002/0471238961.0315191318151905.a01.pub3

Sahota A (2013) Sustainability: how the cosmetics industry is greening up. Wiley. https://doi.org/10.1002/9781118676516.fmatter

Schroeder P, Anggraeni K, Weber U (2019) The relevance of circular economy practices to the sustainable development goals. J Ind Ecol 23(1):77–95. https://doi.org/10.1111/jiec.12732

Schuster F, Throl C (22 Mar 2021) Rezyklate: Wie viel Kunststoff wird wirklich recycelt und was kann jeder tun? [only in German]. Oekotest.de. https://www.oekotest.de/freizeit-technik/Rezyklate-Wie-viel-Kunststoff-wird-wirklich-recycelt-und-was-kann-jeder-tun_11799_1.html

Thüring M, Mahlke S (2007) Usability, aesthetics and emotions in human–technology interaction. Int J Psychol 42(4):253–264. https://doi.org/10.1080/00207590701396674

UN Department of Economic and Social Affairs (UNESA) (2022) The 17 goals. Sustainable Development. https://sdgs.un.org/goals

United Nations Environment Programme (UNEP) (2011) Decoupling natural resource use and environmental impacts from economic growth, a report of the working group on decoupling to the international resource panel (DTI/1388/PA). UNEP/International Resource Panel. http://www.gci.org.uk/Documents/Decoupling_Report_English.pdf

United Nations Environment Programme (UNEP) (2022) Visualising circularity. UNEP circularity platform. https://buildingcircularity.org/

Velenturf APM, Purnell P (2021) Principles for a sustainable circular economy. Sustain Prod Consum 27:1437–1457. https://doi.org/10.1016/j.spc.2021.02.018

VerpackG—Verpackungsgesetz vom 5. Juli 2017 (BGBl. I S. 2234), das zuletzt durch Artikel 2 des Gesetzes vom 22. September 2021 (BGBl. I S. 4363) geändert worden ist [only in German] (2021). https://www.gesetze-im-internet.de/verpackg/

Wakefield-Rann R (2017) More than skin deep: a service design approach to making the luxury personal care industry more sustainable. In: Gardetti MA (ed) Sustainable management of luxury. Springer, pp 211–231. https://doi.org/10.1007/978-981-10-2917-2_10

Wikström F, Verghese K, Auras R, Olsson A, Williams H, Wever R, Grönman K, Kvalvåg Pettersen M, Møller H, Soukka R (2019) Packaging strategies that save food: a research agenda for 2030. J Ind Ecol 23(3):532–540. https://doi.org/10.1111/jiec.12769

Willskytt S (2021) Design of consumables in a resource-efficient economy—a literature review. Sustainability 13(3), Article 3. https://doi.org/10.3390/su13031036

Wilts H, von Gries N (2015) Suffizienz als Geschäftsfeld der Kreislaufwirtschaft [only in German]. uwf UmweltWirtschaftsForum 23(1):41–47. https://doi.org/10.1007/s00550-015-0351-x

Wölfel C, Krzywinski J (2019) Human needs as the crux of the matter in product-service systems development. In: Eriksson Y, Paetzold K (eds) Human behaviour in design. Proceedings of the 2nd SIG conference. Design Society, pp 47–49

World Economic Forum (WEF), Ellen MacArthur Foundation, & McKinsey & Company (2016) The new plastics economy: rethinking the future of plastics. EMF. https://ellenmacarthurfoundation.org/the-new-plastics-economy-rethinking-the-future-of-plastics

Yablonski JI, Mancuso SE (2011) Microbial risks and eco-friendly packaging. In: Formulating, packaging, and marketing of natural cosmetic products. Wiley, pp 179–211. https://doi.org/10.1002/9781118056806.ch11

Zhu Z, Liu W, Ye S, Batista L (2022) Packaging design for the circular economy: a systematic review. Sustain Prod Consum 32:817–832. https://doi.org/10.1016/j.spc.2022.06.005

Can a Circular Economy Lead to Resource Conservation? A Case Study of Long-Term Resource Efficiency Measures in the Small Manufacturing Company

6

Anastassija Konash

1 Introduction

Globalisation, technological advancement, and resource consumption have enabled economic growth and social progress. However, at the same time, the frequency of extreme weather events such as heatwaves, fires, floods, and droughts has increased in recent decades, suggesting that the planetary capacity to sustain human development has been overwhelmed. As a result, people are increasingly looking at ways to transition to more sustainable living (Hedberg et al., 2019).

Circular economy (CE) is a model aiming at decoupling economic growth from resource constraints by keeping materials and resources in the economy for as long as possible, thus minimising waste and virgin resource use. Decoupling economic growth from resource use is expected by redesigning the products and processes and decreasing the material's use in operations and the amount of waste generated. The concept has been around since the late 1960s and has been discussed under several definitions, such as "regenerative design", "industrial ecology", and "cradle to cradle" (Hens et al., 2018). Sustainable development is CE's desired end goal that may be achieved by applying circular economy strategies in social, economic, and environmental dimensions (Lindgreen et al., 2020).

CE goes beyond the conventional "reduce, reuse, and recycle" approach, including repurposing and rethinking materials and repairing, refurbishing, and maintaining products to be cycled back into supply chains. New business models, which allow for shifting from selling products to selling services, have been described (Guldmann and Huulgaard, 2020; Kristoffersen et al., 2020; Lacy et al.,

A. Konash (✉)
Fulbright Coral Sea Scholar Placement, Golisano Institute for Sustainability, Rochester Institute of Technology, 190 Lomb Memorial Drive, Rochester, NY 14623-5608, USA
e-mail: stacey.konash@gmail.com

© The Author(s), under exclusive license to Springer Nature Singapore Pte Ltd. 2024

G. B. Melles and C. Wölfel (eds.), *Design for a Sustainable Circular Economy*, Design Science and Innovation, https://doi.org/10.1007/978-981-99-7532-7_6

2014; Lewandowski, 2016). CE promises to reduce the use of resources in the economy while simultaneously increasing employment opportunities and achieving economic growth (European Commission, 2020). The decrease in resource use and the associated amount of pollution is critical to allow time for pollution mitigation technologies such as carbon capture and carbon storage to mature (Kearns et al., 2021).

However, researchers have identified several limitations to the ability of CE to deliver on its promises of resource conservation (Korhonen et al., 2018). Complex compensatory mechanisms, collectively called rebound effects, come into play to offset efficiency gains and result in higher resource use. The rebound effect is a suite of behaviour and economic changes whose combined effect reduces or eliminates energy and material savings achieved through efficiency measures (Aramendia et al., 2021; Gillingham, 2016).

The rebound effect is sometimes referred to as Jevon's paradox. It occurs when technological progress or governmental policy increases the efficiency of resource use which results in the increased rate of consumption of that resource as the falling cost of use increases demand (Bauer et al., 2009). The effect counteracts the ability of efficiency measures to lower resource consumption and is often ignored by the proponents of efficiency measures and policymakers (Freire-González, 2021).

The role and mechanisms of the rebound effect in the circular economy have been described (Castro et al., 2022; Zink and Geyer, 2017). Moreover, Zink and Geyer (2017) noted that the economic conditions at which circular companies can avoid the rebound effect (matching the prices and taking the customers off the primary producers) were not part of the CE approach being proposed to the companies. This mismatch creates a conflict between some of the currently proposed circularity indicators, such as resource-saving, with other indicators, such as growth and new market creation (Ellen MacArthur Foundation et al., 2015).

The lack of a unified approach to implementing the circular economy on a micro level and the mixed messages regarding the success resulted in the reluctance of many companies to adopt circular economy business models. At the same time, companies are essential agents in the economy and their participation is required if any change is to be achieved practically.

When the adoption of a circular economy with Australian SMEs was discussed in 2020, many owners felt it was prudent to investigate decreasing the environmental impact of their business, but it was not high on their priority list. They valued business survival and profitability for providing their families and employees with a livelihood. Having limited resources to implement changes to their processes, the environmental concerns alone did not provide sufficient reason to embark on the change journey. As a result, it was not clear what steps would be required.

Similar confusion on how to transition to CE was uncovered in the survey of leading companies in Greece, even among large companies (Trigkas et al., 2020). Another study of 286 small and medium companies in France, Belgium, and the UK found that 23% of business owners wanted to see the quantified practical economic benefit of the circular economy, which was followed by incentives (17%) and funding (15%) for the implementation (Fusion, 2014).

This chapter uses the case study on a zero-water, zero-carbon manufacturer's energy efficiency journey to show the efficiency measures' drivers, barriers, and outcomes and to demonstrate to the designers the environment in which companies are making business decisions. Some case study analysis has been published recently (Konash and Nasr, 2022). The interplay between the economic and technical sides of manufacturing the product is discussed in the context of the circular economy rebound as it happened at the company level. The recommendations highlight that profit-making is often incompatible with the overall reduction of environmental impacts.

2 The Case Study

The following illustrative case study is based on evidence gathered during a Full-bright Fellowship by the author based in Rensellaer Polytechnic Institute (RPI) in 2021. Illustrative case studies (Simons, 2009) enable an embedded analysis and reporting of organisational change.

The company was founded in 1977 in New York State. It started as a general-purpose machine shop specialising in manufacturing complex parts through CNC machining and injection moulding. The customer segments included industries such as industrial automation, medical, aerospace, agricultural, electronic, robotics, oil and gas, hardware, plumbing, optics, and other manufacturers. In 2021, the company employed 150 people and had a turnover of $20mln. The company used 40 CNC machines, 30 injection moulding machines, and plastic and metal 3D printers. Materials worked included different qualities of steel, aluminium, brass, copper, and engineering plastics. Other value-added services provided included CAD design, prototyping, CAM programming, assembly, laser welding, laser engraving, inventory management, and outsourced finishing services (grinding, plating, heat treating, and nodizing).

The company started its journey in sustainable manufacturing back in 1990. The electricity supply at the time had stability issues with power outages and surges costing the company tens of thousands of dollars in damaged equipment, labour, and wasted material. The solution suggested by the utility company was to install the transformer for US$100k (cost to the company) that the utility company would own. However, even then, the utility company would not guarantee that the transformer would solve the supply instability. So the company decided to investigate other options and thus started its journey of becoming one of the few zero-carbon manufacturing companies in the world. In 2001, cogenerating heat and power microturbines were installed. They were upgraded in 2015. In 2002, the company became one of the first wind turbine owners in the US. The company upgraded its lighting system in 2007 and again in 2015.

Getting certified for ISO14001 opened up new ways to decrease the environmental footprint. The hydraulic mechanisms on injection moulding machines were substituted for electric. This modification resulted in quieter operations and lower energy requirements. The moulding machines were further modified with glass fibre insulation of the company's design. Installing an absorption chiller allowed cost-effective air-conditioning on the manufacturing floor, thus creating more comfortable working conditions and saving on the dryer for the materials. A pond was constructed next to the facility to provide a process water heat sink and the water for fire system sprinklers. Improved energy management resulted in lower water requirements for cooling and lower overall water consumption by the company. The efforts to improve material efficiency resulted in establishing a new revenue source from selling the company's pre-customer plastic waste as input for other companies. The company invested in collection, preparation, storage, and logistics to enable the sales of its waste stream. The company collaborated with several local charities and not-for-profit organisations to further education in renewable energy and support people affected by the lack of clean water. The company developed and manufactured a water purification product for low-resource settings.

Table 1 shows the connection between CE strategies and relevant company initiatives.

The company started its transformation before the circular economy became prevalent in the public discourse. Nevertheless, the company implemented several strategies compatible with CE by targeting environmental impact. To further analyse the company's fit with CE, the interpretation of the CE concept from Lindgreen et al. (2020) was used (Table 2).

Table 1 The fit between CE strategies and the company's initiatives

Circular economy strategy	Company's initiatives	Comments
Renewable energy	Wind turbine	The energy output is not suitable for the business operation, difficulties with integration into the utility company system
Decrease waste	Sort all waste; return packaging; identify a user for pre-consumer waste	Only one company was interested in using the waste. Additional resources for sorting and storing the materials
Regenerate natural systems	Pond ecosystem established	Space required as well as the redesign of the water-cooling system
Decrease resource use	Energy-efficient measures (lighting, heating, energy generation on-site)	Governmental funding, difficulties with utility company integration
Social impact	The profit-sharing scheme; water purification device for developing economies	Democratic decision-making, multiple stakeholders management

Table 2 Matching CE concept to previous concept and company's achievements

CE concept	Concept explanation	Company's achievement	Pre-CE concept
Value retention	Aim to decouple raw material extraction and growth	Recycling the waste, decreasing water consumption	Sustainable manufacturing
Hierarchical framework	Provides priorities of resource management: reduce, reuse, recycle	Achieved zero-carbon, zero-water manufacturing. It aims at zero waste. Sort waste for internal and external use	Zero-waste manufacturing
Sustainable development	Multidimensional impact: environmentally friendly, economically viable, and socially just	Implemented a profit share scheme, and worked with NGOs on access to clean water	Triple bottom line

Implementing all the initiatives that fit into the CE concept did not fundamentally change the company's business model (Osterwalder and Pigneur, 2010). Key activities and resources remained the same. The sorting of waste and better working conditions became a part of standard manufacturing processes. The cost and revenue structure remained largely the same. The revenue from selling the company's waste plastic did not add substantially to the overall revenue due to the cost of logistics and the limited number of companies interested in purchasing the waste. For example, the company had to rent additional warehouse space to accumulate enough waste material to make shipping viable. The company expanded its network of stakeholders to include energy and material efficiency-promoting organisations and created a new value proposition for its customers: carbon–neutral manufacturing and reporting. However, the new connections did not result in new customers. The company discovered that the sustainability network was more of a community of like-minded people than a customer-recruiting channel.

Many of their usual customers were either unfamiliar with the carbon–neutral manufacturing concept or did not consider it a service for which they were willing to pay. In addition, the larger global companies were not interested in adding the case study company to their suppliers' network due to the complicated supplier evaluation process. Being a contract manufacturer to several OEMs, the company had little influence on the product design. In addition, the recovery of the parts from the final product was also tricky as the company did not have direct access to the product user. When managing the upstream value chain as a small manufacturer, the company had limited influence on its suppliers. For example, although it requested the suppliers to provide ethically sourced materials, the company could not implement any formal system to enforce this request.

This case study demonstrated how the company must work within the limitations and constraints of a small company in the middle of a supply chain.

On the positive side, the company created a reputation for being an innovative and responsible business. The introduction of energy-efficient technologies increased the business's profitability and improved the company's standing in

Table 3 Examples of business case calculations for energy efficiency measures

Project	Scope	Motivation	Projected benefit	Cost	ROI
250kW wind turbine	Install and operationalise the wind turbine	The company needs >3 million kWh/year to operate; 3 million kWh costs $420k/year ($0.14/kWh); the company also pays $$ for taxes and $$ for insurance	Projected energy production = 300,000 kWH +/− 10% per year Electric savings $42,000/year that adds directly to the bottom line	$400,000	9–10 years
Lighting system upgrades	Replace every fixture and ballast plus high bay sodium with new T-8 type fluorescent bulbs and sensors	To improve the light quality on the shop floor by using wider spectrum light and save on electricity, the company needed new reflectors	Total annual electric savings $38,000 per year Staff satisfaction/ improved productivity	$65,000 (decreased to $41,000 via NYSERDA grant and direct federal tax credit)	1.5 year

the community. The company has become an employer of choice for the new generation of workers. The company established itself as a local and global leader in sustainable manufacturing. It educated customers, employees, and the broader community on sustainability and energy efficiency. The company developed reporting metrics for economic benefits derived from resource efficiency. The company's owner credited economic and financial benefits as the main driver for implementing sustainability features.

The examples of calculations that the company uses to build a business case and secure funding for the efficiency measures are presented in Table 3.

3 Waste Reuse and Recycling

The decision on the material used for manufacturing (including any recycled material) lies with the company designing the product. Therefore, the manufacturing company has to follow the material requirements of the designers. Most importantly, the product's design may have to be changed to incorporate any reused material and preserve the performance. Before including any recycled material, an investigation into the trade-offs and impact transfers is required. In addition, the quantity and quality of recycled material vary, increasing the complexity of the input materials in the design process.

However, the first task in the reuse process is to make the material available. The case study company contributed by separating, sorting, shredding, and finding applications for plastic waste. One example of such waste produced by injection moulding is mixed plastic waste. This waste is made during the resin changeover when the machine has produced parts of different kinds of resin. This material contains a mixture of both resins and is called a purge. This material cannot be automatically put back into the company's process as the properties of the purge are not identical to the virgin resin. While unable to reuse it themselves, the company located another plastic parts manufacturer that could utilise this type of mixed waste to improve the quality of their products. Thus, in a circular fashion, the case study company created an example of an industrial ecosystem (Hagnell and Åkermo, 2019), where the waste from one company becomes the feeding stock to another company.

The case-study company's leading service was high-variety low-volume injection moulding manufacturing using high-performing engineering resins. The injection moulding process produced a lower waste volume than CNC manufacturing. Some scrap could be recycled back into the manufacturing process. For manufacturing reuse purposes, post-industrial (or pre-consumer) scrap could be carefully separated, shredded, mixed with virgin material, and fed back into the injection moulding machine. For the case-study company, the customer determined the limit on how much of such pre-consumer plastic waste could be used. The durability, colour pigment, and other factors influenced the amount that can be reused.

4 Circular Economy: Sufficiency and Degrowth

It is important to note that the growth of economic activities, circular included, necessarily leads to increased environmental impact. Population growth, improved living standards, and urbanisation in developing countries drive economic growth that contributes to the global environmental impact. Thus, resource conservation and waste reduction in manufacturing, as opposed to new market creation (Ellen MacArthur Foundation et al., 2015; Victorian Department of Environment Land Water and Planning, 2020), became the primary goal of the circular economy and a means for a more sustainable lifestyle. While material availability issues might slow the introduction of new technologies and the manufacture of new products, climate change due to increased pollution threatens the foundation of societal stability (UN Security Council, 2021). Conserving the resources and the associated pollution is critical to allow time for carbon capture and carbon storage technologies to scale up (Kearns et al., 2021). Until recently, CE resource conservation efforts were focused on resource efficiency initiatives. For example, for energy use, energy efficiency and introducing renewable energy sources were the main strategies for lowering the carbon footprint of manufacturing processes (Thomas et al., 2019). However, as all technologies impact the environment and thus have a limit on their extension (Krier and Gillette, 1985), there is a growing call for sufficiency as an additional necessary strategy for constrainingenvironmental impact of any future energy and material mix (Spengler, 2016).

Concentrating on only the first two strategies (efficiency and renewables) as environmental crisis solutions may lead to the failure to solve the crisis and create undesirable social and ecological consequences (backfire and rebound effect) (Freire-González, 2021). For example, despite substantial technological development in energy efficiency, global energy demand continues to grow at a steady 2% per year (Enerdata, 2021). While more controversial and less developed than energy efficiency and renewable energy, energy sufficiency is a necessary complementary strategy (Muller, 2009; Thomas et al., 2019). All three concepts: energy efficiency, renewable energy, and energy sufficiency, are required to achieve the desired ecological and social outcome of conserving energy (Burke, 2020).

The widespread marketing and promotion of a high-consumption lifestyle as desirable or normal leads to a low social acceptance of sufficiency measures. However, if the contemporary consumption culture remains, CE will fail to change the course of the current unsustainable economic development (Grosse, 2010). For example, the desired outcome of circular economy initiatives originates in the conflict between energy consumption required for human well-being and the adverse environmental and social effects of generating that energy. It is widely understood that energy use and energy services are a means for achieving human well-being and not an end in themselves (Chiao et al., 2011; Smil, 2010). Humans require a minimal level of energy use to meet their basic needs. However, a maximum energy consumption level must also exist. The increase in life quality due to higher energy consumption does not justify the environmental and social harms created during the generation of this extra energy. A recent review has found that a high level of human well-being could be supported by a relatively low amount of energy (Burke, 2020).

Given that the climate change emergency is mainly driven by the changes in the atmosphere resulting from greenhouse emissions, reducing energy consumption, which is responsible for most greenhouse gas emissions worldwide, seems like a good starting point and a priority. However, the strong historical link between energy consumption and economic growth questions whether continued economic growth is compatible with energy conservation targets (Aramendia et al., 2021).

It is possible that global warming is inseparable from economic growth, and reducing it may lead to reduced economic growth as measured by GDP (Lang and Gregory, 2019). Suppose a more significant decoupling of energy consumption from economic growth is not achieved. In that case, it will be necessary to rapidly scale up low-carbon energy supply, carbon capture, and storage technologies to meet energy demand and prevent catastrophic global warming. These low and negative-emission technologies have limitations, including large-scale investment, extensive land use, and significant lead times. Thus, expanding them will be politically challenging and will take time (Aramendia et al., 2021).

On the other hand, an overemphasis on technical matters of energy generation, transition to renewable energy sources, and carbon capture may prevent society from questioning the necessity of high energy consumption. Therefore, it is crucial to consider the societal structures that drive high levels of energy use to ensure the application of technology will achieve the desired energy conservation goals (Toulouse et al., 2019). There is growing momentum among scholars and activists to advocate "degrowth," a critique of capitalist economic development that supports the shrinking of production and consumption, reorienting societies to use fewer natural resources and to live more sustainably (e.g. Schröder et al., 2019).

Recently, there has been an increased emphasis on sufficiency strategies—refuse, rethink, and reduce in CE literature (Bocken et al., 2022). A similar shift was observed previously in the energy sector regarding energy conservation efforts through a sufficiency strategy (Burke, 2020; Krier and Gillette, 1985; Muller, 2009; Spengler, 2016; Thomas et al., 2019). The sufficiency model is closely related to degrowth concepts (Buch-Hansen and Carstensen, 2021). Both models recognised that all human activities, including economics, were subject to physical laws; thus limitless growth in a limited environment was impossible (Cosmea et al., 2017). However, unlike degrowth, the sufficiency strategy allows for growth as long as this growth is environmentally sustainable (Bocken and Short, 2016). Several reports described examples of companies looking to adopt degrowth or sufficiency models (Bocken et al., 2022; Nesterova, 2021) in similar terms. Both reports indicate that "no growth" is impossible for new businesses. As the founder of a degrowth startup puts it: ".. you cannot build a company that doesn't grow because that isn't a good experience for the people in the business. We must address how we grow, dematerialise growth, and create businesses with less drawdown on natural capital. The pace of growth most VCs (venture capitalists, SK) have come to expect is not sustainable for the planet or the people in the business" (Webb, 2022).

The discussion with the case study company owner revealed that the "no growth" strategy had widespread opposition among stakeholders in the current economic situation. It went against the expectations of banks, employees, and customers. Banks require growth to secure a loan, which is essential for business continuity and innovation. The employees who partook in larger profits through the profit-share scheme also needed growth. It demonstrated the difficulties facing CE approaches that aim to improve all three areas of human activities: social, environmental, and economical. It again highlighted that in an environment with limited resources, growth in one place came at a cost and deterioration in another area (Moreau et al., 2017).

As the owner was motivated to do the right thing for the environment and minimise the business's ecological footprint, temporary energy conservation was achieved with constant investment in innovative technologies. However, the company's expansion resulted in other environmental impacts such as higher land use (pond, wind turbines), construction of the new building, and higher consumption potential for the employees through profit sharing. Therefore, even environmentally-driven business decision-makers cannot achieve the desired decrease in the company's footprint. Significant structural socio-economic changes are needed to enable individual companies to consider sufficiency strategies along with efficiency measures. Without sufficiency-directed policies and regulations (Thomas et al., 2019), companies are trapped in the "business as usual" model. As a result, they are limited to efficiency and renewable energy strategies, which fail to provide long-term resource conservation outcomes, as demonstrated here.

The interview with other business owners in the same geographic region revealed strong scepticism about the ability of efficiency measures and renewable energy to provide a complete solution for climate change (unpublished data). This observation suggests that highlighting best resource efficiency practices without policies and regulations limiting the use of resources is unlikely to inspire the necessary change in the manufacturing sector. On the contrary, concentrating on resource conservation while curbing consumption as the primary goal of circular economy activities will increase CE potential to enable long-term sustainable development (Basiago, 1998; Moreau et al., 2017).

5 Designers as Change Drivers: Conviviality

The main appeal of circular economy is the ability to design reduced use of material and energy, thus enabling the goal of equity in harmony with the environment. The resulting society models proposed include "slow living", minimalist living, and localised self-sufficient communities.

For the designers to support such living, the concept of "conviviality" was introduced and developed (Lizarraldea and Tyl, 2018). A Low-tech and sufficiency approach presents a potential solution to industrial and global technological solutions. The designers' choices of implemented technology and materials play an essential role in socio-political, economic, and cultural outcomes. In one example, the promotion of small-scale, distributed technology production and end-of-life systems (i.e. localisation) enhance autonomy and equity while providing access and knowledge to produce and maintain technologies (Lizarraldea and Tyl, 2018). Another conviviality criterion—frugal material use—challenges the designers to create less complex products with locally-sourced materials, which again improves the autonomy and stability of the supply chain.

The primary strategy to address the circular economy rebound effect should be to ensure that the secondary market targets the same consumers at the same price point. This outcome could be achieved by providing equivalence in quality between the primary and secondary products and educating consumers to remove the prejudices associated with secondary products.

The second strategy to avoid the rebound effect from circular activities is limiting those activities to markets with low product price sensitivity controlled by a few companies. In this case, it would be possible to ensure that the increase in the supply from the introduction of secondary products does not lower the price or increase the demand for the product. For example, large agricultural machinery could be one such market. Conversely, in markets where the prices cannot be controlled and the demand/price sensitivity is high, the increase in supply would lower the price, increasing the demand.

Both strategies should result in effective 1:1 substitution of the primary products by the secondary products. However, such a substitution would mean the shrinking of the primary market for the manufacturer, which goes against the interests of multiple stakeholders in the current economic environment. Thus, it is unlikely that any company would target either of the above strategies to stay competitive in the current economic conditions.

6 Conclusion

While analysing the case study's business model, it was found that little to no change in the business model was necessary to implement efficiency-based sustainability measures. This outcome suggests that changing the business model is unnecessary for a company to transition to a circular economy. A detailed discussion of the energy, water, and material efficiency measures was presented, with the drivers and challenges for each. Implementing resource efficiency initiatives was mainly driven by the strong moral beliefs of the owner and the desire to differentiate and establish the company as an innovative and responsible leader. Integrating energy generation technologies with the utility grid was the biggest challenge. Sustained innovation and entrepreneurship were the essential enablers.

Two strategies for the designers to avoid the rebound effect were presented. However, it seems unlikely they can be implemented in the current economic conditions. Therefore, there is a need to find new forms of interpretation and intervention to confront environmental crises and challenge corporate visions of the circular economy. The most urgent priority is to challenge entrenched corporate and societal views about growth. Current circular economy policies fail to challenge the capitalist imperative for growth, glossing over "reduction" among the *R*s of the circular economy. However, on a deeper level, which goes beyond the limits of the circular economy, there is a need to tackle questions about values, inequality, and future generations.

Acknowledgments Anastassija Konash gratefully acknowledges financial support for this research by the Fulbright U.S. Scholar Program, which the U.S. Department of State and the Australian-American Fulbright Commission sponsors. This chapter is solely the author's responsibility and does not necessarily represent the official views of the Fulbright Program, the Government of the United States, or the Australian-American Fulbright Commission. The author thanks Professor Nabil Nasr for his support during the fellowship and the case study data collection.

References

Aramendia E, Brockway PE, Pizzol M, Heun MK (2021) Moving from final to useful stage in energy-economy analysis: a critical assessment. Appl Energy 283:116194. https://doi.org/10.1016/j.apenergy.2020.116194

Basiago AD (1998) Economic, social, and environmental sustainability in development theory and urban planning practice. Environ 19:145–161. https://doi.org/10.1023/A:1006697118620

Bauer D, Papp K, Polimeni J, Mayumi K, Giampietro M, Alcott B (2009) John Polimeni, Kozo Mayumi, Mario Giampietro & Blake Alcott, the Jevons Paradox and the myth of resource

efficiency improvements. Sustain Sci Pract Policy 5:48–54. https://doi.org/10.1080/15487733.2009.11908028
Bocken NMP, Short SW (2016) Towards a sufficiency-driven business model: experiences and opportunities. Environ Innov Soc Transitions 18:41–61. https://doi.org/10.1016/j.eist.2015.07.010
Bocken NMP, Niessen L, Short SW (2022) The sufficiency-based circular economy—an analysis of 150 companies. Front Sustain 3:1–18. https://doi.org/10.3389/frsus.2022.899289
Buch-Hansen H, Carstensen MB (2021) Paradigms and the political economy of ecopolitical projects: green growth and degrowth compared. Compet & Chang 25:308–327. https://doi.org/10.1177/1024529420987528
Burke MJ (2020) Energy-suffciency for a just transition: a systematic review. Energies. https://doi.org/10.3390/en13102444
Castro CG, Trevisan AH, Pigosso DCA, Mascarenhas J (2022) The rebound effect of circular economy: definitions, mechanisms and a research agenda. J Clean Prod 345:131136. https://doi.org/10.1016/j.jclepro.2022.131136
Chiao RY, Cohen ML, Leggett AJ, Harper CLJ, Phillips WDP (eds) (2011) Visions of discovery: new light on physics, cosmology, and consciousness. Cambridge University Press, Cambridge
Cosmea I, Santosa R, O'Neill WD (2017) Assessing the degrowth discourse: a review and analysis of academic degrowth policy proposals. J Clean Prod 49:321–334. https://doi.org/10.1016/j.jclepro.2017.02.016
European Commission (2020) Communication from the commission to the European parliament, the council, the European economic and social committee and the committee of the regions a new circular economy action plan for a cleaner and more competitive Europe. Brussels
Ellen MacArthur Foundation, McKinsey Center for Business and Environment, Stiftungsfonds Für Umweltökonomie und Nachhaltigkeit (SUN) (2015) Growth within: a circular economy vision for a competitive europe. Ellen MacArthur Foundation
Enerdata (2021) Total energy consumption [WWW Document]. Glob Energy Stat Yearb 2021
Freire-González J (2021) Governing Jevons' Paradox: policies and systemic alternatives to avoid the rebound effect. Energy Res Soc Sci 72. https://doi.org/10.1016/j.erss.2020.101893
Fusion (2014) How to shift towards the circular economy from a small and medium business perspective: a guide for policy makers. 1–23.
Gillingham K (2016) Rebound effects. In: The new Palgrave dictionary of economics. Palgrave Macmillan UK, London, pp 1–12. https://doi.org/10.1057/978-1-349-95121-5_2875-1
Grosse F (2010) Is recycling "part of the solution"? The role of recycling in an expanding society and a world of finite resources. Sapiens 3
Guldmann E, Huulgaard RD (2020) Barriers to circular business model innovation: a multiple-case study. J Clean Prod 243. https://doi.org/10.1016/j.jclepro.2019.118160
Hagnell MK, Åkermo M (2019) The economic and mechanical potential of closed loop material usage and recycling of fibre-reinforced composite materials. J Clean Prod 223:957−968. https://doi.org/10.1016/j.jclepro.2019.03.156
Hedberg A, Šipka S, Bjerkem J (2019) Creating a digital roadmap for a circular economy. Discuss Pap-Sustain Prosper Eur Program 24
Hens L, Block C, Cabello-Eras JJ, Sagastume-Gutierez A, Garcia-Lorenzo D, Chamorro C, Herrera Mendoza K, Haeseldonckx D, Vandecasteele C (2018) On the evolution of "cleaner production" as a concept and a practice. J Clean Prod 172:3323–3333. https://doi.org/10.1016/j.jclepro.2017.11.082
Kearns D, Liu H, Consoli C (2021) Technology readiness and costs of CCS. Global CCS Institute
Konash A, Nasr N (2022) The circular economy and resource use reduction: a case study of long-term resource efficiency measures in a medium manufacturing company. Clean Prod Lett 3:100025. https://doi.org/10.1016/j.clpl.2022.100025
Korhonen J, Honkasalo A, Seppälä J (2018) Circular economy: the concept and its limitations. Ecol Econ 143:37–46. https://doi.org/10.1016/j.ecolecon.2017.06.041
Krier JE, Gillette CP (1985) The un-easy case for technological optimism. Mich Law Rev 84:405. https://doi.org/10.2307/1289008

Kristoffersen E, Blomsma F, Mikalef P, Li J (2020) The smart circular economy: a digital-enabled circular strategies framework for manufacturing companies. J Bus Res 120:241–261. https://doi.org/10.1016/j.jbusres.2020.07.044

Lacy P, Keeble J, McNamara R, Rutqvist J, Haglund T, Cui M, Cooper A, Pettersson C, Eckerle K, Buddemeier P, Sharma A, Senior T (2014) Circular advantage: innovative business models and technologies to create value in a world without limits to growth. Accenture Strategy

Lang PA, Gregory KB (2019) Economic impact of energy consumption change caused by global warming. Energies 12:1–29. https://doi.org/10.3390/en12183575

Lewandowski M (2016) Designing the business models for circular economy-towards the conceptual framework. Sustain 8:1–28. https://doi.org/10.3390/su8010043

Lindgreen ER, Salomone R, Reyes T (2020) A critical review of academic approaches, methods and tools to assess circular economy at the micro level. Sustain 12. https://doi.org/10.3390/su12124973

Lizarraldea I, Tyl B (2018) A framework for the integration of the conviviality concept in the design process. J Clean Prod 197:1766–1777. https://doi.org/10.1016/j.jclepro.2017.03.108

Moreau V, Sahakian M, van Griethuysen P, Vuille F (2017) Coming full circle: why social and institutional dimensions matter for the circular economy. J Ind Ecol 21:497–506. https://doi.org/10.1111/jiec.12598

Muller A (2009) Sufficiency–does energy consumption become a moral issue? IOP Conf Ser Earth Environ Sci 6:262003. https://doi.org/10.1088/1755-1307/6/6/262003

Nesterova I (2021) Small firms as agents of sustainable change. Futures 127:102705. https://doi.org/10.1016/j.futures.2021.102705

Osterwalder A, Pigneur Y (2010) Business model generation: a handbook for visionaries, game changers, and challengers. Strategyze. ed. Wiley-VCH Verlag.

Schröder P, Bengtsson M, Cohen M, Dewick P, Hoffstetter J, Sarkis J (2019) Degrowth within–aligning circular economy and strong sustainability narratives. Resour Conserv Recycl 146:190–191. https://doi.org/10.1016/j.resconrec.2019.03.038

Simons H (2009) Listen, look, document: methods in case study research. In: Case study research in practice. pp 43–67. https://doi.org/10.4135/9781446268322.n3

Smil V (2010) Science, energy, ethics, and civilization. In: Visions of discovery: new light on physics, cosmology, and consciousness. pp 709–729

Spengler L (2016) Two types of 'enough': sufficiency as minimum and maximum. Env Polit 25:921–940. https://doi.org/10.1080/09644016.2016.1164355

Thomas S, Thema J, Brischke LA, Leuser L, Kopatz M, Spitzner M (2019) Energy sufficiency policy for residential electricity use and per-capita dwelling size. Energy Effic 12:1123–1149. https://doi.org/10.1007/s12053-018-9727-4

Toulouse E, Bohnenberger K, Bierwirth A, Leuser L, Sahakian M, Lorek S (2019) Energy sufficiency: how can research better help and inform policy-making?. In: Eceee summer study proceedings. pp 331–339

Trigkas M, Karagouni G, Mpyrou K, Papadopoulos I (2020) Circular economy. The Greek industry leaders' way towards a transformational shift. Resour Conserv Recycl 163. https://doi.org/10.1016/j.resconrec.2020.105092

UN Security Council (2021) Climate change 'biggest threat modern humans have ever faced',world-renowned naturalist tells security council, calls for greater global cooperati

Victorian Department of Environment Land Water and Planning (2020) Recycling Victoria: a new economy

Webb B (2022) Early majority: fashion's first degrowth brand [WWW Document]. Vogue

Zink T, Geyer R (2017) Circular economy rebound. J Ind Ecol 21:593–602. https://doi.org/10.1111/jiec.12545

Circular Design for a Transition to a Sustainable Circular Society: Defining a New Profession

7

Gavin Brett Melles and Anne Velenturf

1 Introduction

In recent years, there has been increasing global enthusiasm for circular economy (CE) as the approach of choice for businesses, industries, and governments to achieve continued economic growth and overcome the challenge of decoupling growth from resource use (D'Amato et al. 2017; Ellen MacArthur Foundation 2013; Fletcher and Rammelt 2017; Ward et al. 2016). The CE model builds on a history of ideas about minimising resource use on an ecologically bounded 'spaceship earth' (Boulding 1966; Crocker 2018), through cradle-to-cradle product design (McDonough and Braungart 2010), a performance economy[1] based around services rather than ownership (Stahel 2010) and other CE and sustainability strategies (Haas et al. 2020; Korhonen, Nuur, et al. 2018; Reike et al. 2018; Stahel 2020; Winans et al. 2017).

Governments have translated circular enthusiasm into national strategies and roadmaps, albeit with varying scope and intent (Asia-Pacific Economic Cooperation 2020; Circular Economy Initiative Deutschland 2021; Poulton and Lyne 2009; Price Waterhouse Cooper 2019; Schandl et al. 2021). CE narratives offer an appealing rationale for business opportunity and green growth in an era of net zero

[1] In fact, Stahel (2020) distinguishes a circular economy and performance economy in that only the latter is a consistent systematic implementation of the former idea.

G. B. Melles (✉)
Faculty of Architecture and Industrial Design, Swinburne University, Melbourne 3122, Australia
e-mail: gmelles@swin.edu.au

A. Velenturf
School of Civil Engineering, University of Leeds, Leeds LS2 9JT, England
e-mail: A.Velenturf@leeds.ac.uk

© The Author(s), under exclusive license to Springer Nature Singapore Pte Ltd. 2024

G. B. Melles and C. Wölfel (eds.), *Design for a Sustainable Circular Economy*,
Design Science and Innovation, https://doi.org/10.1007/978-981-99-7532-7_7

ambitions (Black et al. 2021; Corvellec et al. 2021; Haucke 2018; Saidani et al. 2017; Temesgen et al. 2019). Despite the enthusiasm, consumption-based evaluation of global circular implementation, based on material footprint (e.g. Wiedmann et al. 2015), shows static or declining progress over the last five years (Platform for Accelerating the Circular Economy (PACE) 2020).[2] Reasons for this gap between circular rhetoric and a reality of negative material and emission feedback are multiple, including excessive reliance on waste and recycling (Valenzuela and Böhm 2017; Vonk 2018), offshoring of material footprint (Horvath et al. 2019; Wiedmann et al. 2015), and many other issues. Hence, there are calls to move beyond the mainstream model, including with respect to circular design (Moreno et al. 2016; Schroeder et al. 2019).

CE strategies, especially recycling (Allwood 2014; Islam and Huda 2019), appear to encourage rather than discourage increased consumption, resource use and emissions, through the so-called rebound effect (Figge and Thorpe 2019; Hobson 2021; Makov and Vivanco 2018; Zink and Geyer 2017), and the false belief that recycling infrastructure and processes are effective (Binet et al. 2019). More generally, circularity is promoted in an era where dominant energy sources remain fossil fuel-based (Corvellec et al. 2021; Jackson and Victor 2020; Kothari et al. 2014; Twomey and Washington 2016). The diversity of definitions and implementations included under the broad umbrella of CE has also been criticised, particularly with respect to increasing greenwashing potential (Corvellec et al. 2021; Figge and Thorpe 2019; Geissdoerfer et al. 2017; Holzinger 2020; Homrich et al. 2018; Kirchherr et al. 2017; Korhonen, Honkasalo, et al. 2018; Makov and Vivanco 2018; Reike et al. 2018; Temesgen et al. 2019). These and other reasons, suggests that circularity for circularity's sake (Harris et al. 2021) rather than a sustainable transition is being promoted.

The differences between the green growth narrative of circular economy (Hickel and Kallis 2020; Wanner 2015), and environmental, social and economic progress towards sustainable development have become increasingly apparent (Alonso-Almeida et al. 2020; Blum et al. 2020; Camilleri 2018; Corona et al. 2019; Desing et al. 2020; Haupt and Hellweg 2019; Johansson and Henriksson 2020; Reike et al. 2018; Schroeder et al. 2019; Velenturf and Jopson 2019; WEF 2020; Whalen and Whalen 2020). In an effort to rescue the impetus of circularity, reconnect this to sustainability, and distinguish modest reform from a circular transformation (Reike et al. 2018), a movement driven by the goal of enabling a circular society is becoming popular (Fan et al. 2019; Jaeger-Erben et al. 2021; Kayikci et al. 2021; Leipold et al. 2021; Melles 2021; Ralph Boch et al. 2020; van der Velden 2021; Velenturf and Jopson 2019; Velenturf and Purnell 2021; Wu et al. 2022). This transformist movement is supported by a holistic circular discourse with closer

[2] The Circularity Gap project measures progress using material footprint indexes (MFI) data, including proxies where necessary, for a consumption account of progress that for a nation as a whole and the materials required to support specific lifestyles.

links to sustainable development considerations (Bauwens et al. 2020; Calisto Friant et al. 2020a, b; De Angelis and Ianulardo 2020; Melles et al. 2022; Twomey and Washington 2016; van den Bergh 2020). Building on an earlier definition of a resource-circulating society (Komiyama and Takeuchi 2006), the circular society vision and the sustainable circular economy principles driving this are becoming the preferred term for a range of stakeholders looking beyond the mainstream narrative.[3]

2 A Plurality of Circular Discourses

The mainstream circular economy narrative is not the only circular discourse on offer. Several scholars with political economy lenses have examined the emergence of any conflict between different accounts of circularity and also the existence within the circular economy of hybrid mixes of narratives. Referring to their work on CE discursive differences as an outcome of their work on the UK Resource Recovery from Waste programme (RRfW), Velenturf and Purnell (2021) identified a continuum 'from resource efficiency, improving existing practices, and weak sustainability on the one hand ... to resource productivity and strong sustainability on the other hand, requiring radical changes to resource use in our society' (2021, 1443). Ortega-Alvarado et al. (2021) founded a range of competing discourses about waste, consumption and sharing economy under the banner of circularity in Norway, while Johanssen and Henriksson (2020) founded weak and strong circularity discourses reminiscent of the circular economy and society distinction. For Australia, Melles (2021) finds mainstream and more holistic circular discourses competing to define the transition in that country, while Friant et al. (2022) articulate a similar account for the Netherlands of circular discourses varying between technocentric and transformational.

To address weaknesses in representing the full range of systemic economic, materials, energy and other challenges to a sustainable CE, Friant et al. (Calisto Friant et al. 2020a, b) develop a typology of circularity discourses,[4] particularly focussed on the contrast between mainstream CE and a circular society account. A central difference is that a reform of capitalism and new economic thinking, respect for ecological boundaries and prosperity for all through social innovation and new business models is necessary for circular society. Thus, Doughnut Economics (Raworth 2017), which argues for an economy based on market, household, government and commons within ecological limits, is a circular society position. Thus, the circular society agenda of socio-political and economic change and vision integrates and expands on the CE focus on material efficiencies and relevant business models.

---

[3] See this Dutch consortium https://ewuu.nl/en/research/circular-society/.

[4] Discourses are narratives that circulate and justify practices (Hardy and Thomas 2015).

3 Circular Discourses and Design

Everyone designs who devises courses of action aimed at changing existing situations into preferred ones (Simon 1996).

The mainstream technicist and holistic circular society discourses offer different opportunities for a sustainable circular economy and design. As noted above, mainstream CE identifies one of its main strategies as designing out waste and pollution (EMF 2015), and industrial and product design strategies are often proposed to include recycled content and waste, as well as adopting eco-efficiency approaches. However, sustainable circular design principles (below) ask us to avoid waste altogether rather than designing it by treating waste and standard recycling as inevitable inputs, and there is far greater scope for design practices with respect to circular society goals (Ralph Boch et al. 2020). This has been recognised by some scholars who have suggested that circular design should expand its remit beyond mainstream industrial and product design considerations (e.g. Moreno et al. 2016). In the following paragraphs, we give a brief excursion into the expanding role of design and compare narrow and broader conceptions of circular design. This is a prelude to presenting sustainable CE principles as the basis for a new definition of sustainable circular design.

An early critique of the unsustainability of expert industrial design was the work of Viktor Papanek—Design for the Real World (Papanek 1971). Responses to Papanek's criticism gave rise over time to the development of social design and the engagement of design in development (e.g. Kumar et al. 2016) and equity issues across the globe (Melles et al. 2011). As a result, design methods and processes, including co-design as part of the new landscapes of design (Sanders and Stappers 2008), have diffused into many social and sustainability domains (Boylston 2019). Thus, in addition to designing out waste and pollution in an industrial and business context, design can be deployed in designing social and political futures (Earley 2017; Fry 2009; Hales 2013; Wastling et al. 2018).

For circular economy, design has been increasingly highlighted as a catalyst (Andrews 2015; Moreno et al. 2016; van Dam et al. 2020). Firstly, existing discussion and materials on circular design include a diverse array of well-known eco-design methods and principles, e.g. design for manufacture, life-cycle analysis (LCA), cradle to and thinking, cradle to cradle design (McDonough and Braungart 2010), and design for recycling, etc. (den Hollander et al. 2017). A good example of the new synthesis is the Circular Design Guide[5]—a joint initiative of EMF and IDEO design agency. Current discussions of circular design typically include deliberations on the circular design of product service systems (Halstenberg and Stark 2019) and business models (Saidani et al. 2017). In addition, proposals for a new circular design curriculum will build on existing practices and knowledge from design for sustainability (Moreno et al. 2016).

[5] Methods (circulardesignguide.com).

Secondly, circular design within circular economy narratives sometimes extends its operational boundary to help define circular business models, including product service systems (McAloone and Pigosso 2018). Hence, there is a need to implement a design for product and service integrity agenda, through employing all the so-called R-strategies from refuse to recycle (den Hollander et al. 2017; McAloone and Pigosso 2018). Other proposals also identify a range of new communicative, e.g. storytelling and strategic competencies for circular design that extend into diffuse design space (Sumter et al. 2020). Thus, an expanded role for industrial design practices to influence production, consumption, policy and education has been proposed recently (van Dam et al. 2020). Hence, this wider sphere of influence already suggests an expanded definition of circular design is the order of the day.

Indeed, policy, regulation, standards and multiple other mechanisms and actors must create an environment in which mainstream industrial design practices would make sense and in turn reinforce a sustainability transition (Allwood 2014; EEA 2019). Thus, circular design discussions occasionally allude to broader societal, economic, and environmental aspects (Bocken et al. 2016; Lofthouse and Prendeville 2018; Moreno et al. 2016). Other more holistic accounts of circular design focussed on its human-centred potential (Lofthouse and Prendeville 2018) offer some guidance on principles and knowledge requirements. This application of more diffuse design thinking and practices is consistent with design for social innovation (Kumar et al. 2016; Manzini 2015), social business models (Burkett 2013), policy design (Howlett 2020; Huybrechts et al. 2017) and systems based social design (Boylston 2019).

Thus, beyond a circular economy vision of technology innovation, a more diffuse circular design focuses on multi-stakeholder strategic re-design of new institutions for a sustainable circular society (Goodin 1996; Hobday et al. 2012; Huybrechts et al. 2017; Ralph Boch et al. 2020). Such an agenda expands the remit of design from technical to social innovation and from narrow industrial design expertise to more diffuse design inputs to social change, as Manzini (2015) has identified. Our chapter takes this agenda up and links it to sustainable circular economy principles as well as the circular society discourse and agenda.

4 Multi-stakeholder and Multi-level Sustainable Circular Design Principles

A systems account of the interactions between social and technical innovation and multi-stakeholder institutional reform is required to explain how sustainability transitions can happen. Systems thinking is one of the overarching principles in this transition, and places emphasis on identifying the feedback and interactions among the variables in the systems (Meadows 2008; Sterman 2000), for example, how circular economy activities lead to a rebound in production outputs (Zink and Geyer 2017) or how recycling may 'surprisingly' increase consumption (Fitch-Roy et al. 2019). Principles are also required to strategically influence this transition, as articulated in multi-level sustainability transitions theory (Loorbach et al. 2017).

Multi-level transitions theory (MLP) meanwhile envisions the change process towards a new socio-technical regime—a circular society or *only* a circular economy—as the product of multi-stakeholder and multi-level innovations in policy and practice (Geels 2011; Kanda et al. 2020; Loorbach et al. 2017; Rauschmayer et al. 2015). Consistent with such an approach, we require a set of principles that can encourage the multi-level technical and socio-economic and political changes to encourage a just transition to a circular society. Below, we suggest sustainable circular economy principles and offer a detailed breakdown of the interdependent strategies needed to achieve this transition.

Velenturf and Purnell (2021) outline a set of ten principles that highlight how to mobilise communities private sector, and the government to develop circular society solutions for specific contexts. Acknowledging sustainable development concerns, it is a model that combines circular economy and society discourses and considerations (Calisto Friant et al. 2020a, b; D'Amato et al. 2017; Jaeger-Erben et al. 2021: Reike et al. 2018). The framework provides scope for broad ranging technical and social innovation through private, public and civic sectors. Their circular design outline (principle 3) argues for far more than the usual re-design of products; products and the materials that they are made of are embedded in supply chains, wider systems of production and consumption, society and the environment. This is calling for a system-wide transformation of industrial systems and society, consistent with the industrial ecology thesis but also beyond (Saavedra et al. 2018). The example (principle 3) of their design for circularity outline illustrates the holistic scope of their proposal.

> Design, select and transform industrial systems, supply chains, materials and products, using "R-ladders" and whole-system assessments of solutions to optimise stocks and the degree of closing loops of resource flows, minimising raw material extraction and waste generation, optimising value generated for people, and enabling reintegration of materials into natural biogeochemical processes at end-of-use, through continuous processes nurturing sustainable solutions, through innovation, and phasing out unsustainable practices, through exnovation, to implement and maintain a sustainable circular society.

Links between the principles are articulated in the other principles. Thus, principles 1–4—reduced resource flows, decoupling of prosperity from material use and consumption through sufficiency and efficiency approaches, circular design (outline above), and circular business models for social, environmental, and economic value and impact set the scene for a modified economic model. Meanwhile principles 5–9 address more radical socio-political changes and mechanisms associated with circular society—consumption transformation, citizen participation, etc. There is an allusion to sustainability transition in principle 6, albeit with multiple possible outcomes, as indicated by principle 8. Changes must be enabled by participatory design and a return to strong sustainability as the foundation of politico-economic change. These radical proposals for designing more preferred socio-economic situations require multi-stakeholder, multi-level processes of engagement and outcome. Principle 10 advocates for system analysis and 'redesign' in support of continuous learning and evaluation of transformation pathways. Achieving these aims will require a coordinated multi-level and multi-stakeholder engagement, including key political, economic and socio-cultural changes towards a new circular socio-technical regime, as proposed in sustainability transitions theory (Geels 2011).

Table 1 lists the principles (altered in some cases), provide a concise outline, and then identify core concepts and approaches which are alluded to in these. In some cases, the concepts alluded to derive from the original paper, while for others, relevant concepts and approaches were added. For the latter, illustrative references are provided and add a column with links to existing design strategies and approaches. The lists and examples are not intended to be exhaustive, there is significant overlap between principles and concepts, hence design strategies could be placed in more than one category and the table is intended to prompt the reader to consider the scope of the new landscapes of design (Sanders and Stappers 2008).

Principle	Short definition	Existing related concepts	Design roles
Nature positive economy	Material extraction rates and energy generation for production and consumption balanced by return to environment, within the planet's carrying capacity	Nature-based solutions (Seddon et al. 2021). Ecosystem stewardship (Chapin et al. 2010). Bioeconomy (D'Amato et al. 2017)	Nature positive design (Birkeland 2022)
Reduce and decouple resource use	Progress is decoupled from unsustainable material use through a focus on efficiency, sufficiency, and dematerialisation	Material circularity (Wiedmann et al. 2015), consumption-based circular assessment (Brown et al. 2018). Sufficiency-driven business models (Bocken and Short 2020)	Eco-efficient design (Ljungberg 2007). Cradle-to-Cradle Design (McDonough and Braungart 2010)

(continued)

(continued)

Principle	Short definition	Existing related concepts	Design roles
Design for circularity	Transform industrial systems, supply chains, materials, and products, using "R-ladders" and whole-system assessments of solutions (P10)	Industrial symbiosis (Lifset and Graedel 2015). Industrial ecology. Waste hierarchy. Circular supply chains (Bressanelli et al. 2019). Life cycle assessment (Unep 2003). Complex value assessment (Iacovidou et al. 2017). Exnovation (Fossati et al. 2022). Sustainable Supply Chains (Smith 2008)	Circular product design (Sumter et al. 2018). Life cycle oriented design (Aurich et al. 2006). Sustainable Circular Design (Moreno et al. 2016)
Sustainable circular business models	Governance enables business models to internalise social and environmental costs of materials and products into their prices	Sustainable circular business models (Antikainen and Valkokari 2016; Bocken et al. 2020)	Design sustainable circular business models (Lewandowski 2016). Designing social business models (Burkett 2013)
Transform consumption practices	Systems of provision enable sufficiency-oriented, demand-driven resource use and more sharing, service, and experience-based consumption	Performance-based economy (Stahel 2010). Post-capitalism consumption (Hobson and Lynch 2016). Sufficiency (Lamberton 2005). Systems of provision	Social enterprise (co)design (Selloni and Corubolo 2017), experience-based design, sustainable product service design (Vezzoli et al. 2017)
Multi-stakeholder social business and innovation	Participatory social innovations bring people, business and policy makers together across system levels	Commons collective action (Ostrom 1990). Social enterprise (Teasdale 2012). Social innovation (Mulgan 2010)	Design for social innovation (Manzini 2015), Co-design for social innovation (Britton 2017)

(continued)

(continued)

Principle	Short definition	Existing related concepts	Design roles
Coordinated multi-level policy and practice	Coordinated implementation of circular economy strategies and actions with societal actors across scales at key intervention points	Sustainability intermediaries (Kivimaa et al. 2019). Circular Governance (Ddiba et al. 2020). Multi-level Sustainability Transitions (Loorbach et al. 2017). Circular Policy (McDowall et al. 2017). PESTLE analysis (Mishra et al. 2019), Participatory Situational Analysis (Koutra 2010)	Future scenario design (Kishita et al. 2016), low-carbon scenario co-design (Shaw et al. 2009), Context analysis
Promote diversity and flexible solution implementation	A plurality of perspectives and local solutions for circular economy and a culture of knowledge exchange and learning across society for resilient circular economy processes	Resilience thinking and practice (Biggs et al. 2012). Scenario planning and design (Kahane 2012). Community participation (Sanoff 2005). Participation process management	Participatory social design (Ralph Boch et al. 2020). Participatory action research
Political economy for prosperity and well-being	Move from short-term GDP focus to long-term prosperity, well-being and environmental quality as goals	Well-being and prosperity focus (Jackson 2009). Doughnut Economics (Raworth 2017). Strong sustainability (Schröder et al. 2019). Multi-dimensional prosperity (Sands 2015). Multi-dimensional value	Policy co-design (McGann et al. 2018)
System design and assessment	Systems thinking approaches to the design and evaluation of circular proposals and transition processes	Planetary boundaries (Rockström and Steffen 2009), Systems thinking (Meadows 2008). Precautionary principle; Resilience thinking (Folke et al. 2010) Complex value assessment	Systems thinking in design (Mononen 2017). Complex value assessment

Table 1 expanded sustainable circular design R-strategies

R0 refuse	Sufficiency-based demand for products and exnovation of unsustainable products. The design and marketing of products and services will be refused where they have unsustainable outcomes of material choices and energy, and do not match the other R-strategy criteria, within the boundaries of a just society
R1 rethink	Rethink the economy, business and industry, and the role of public goods and markets. This entails new business and service models, social innovation, and consistency with a well-being focus. Rethinking enables the other r-strategies and is a result of the participatory processes listed above. Overall rethinking how demand for resource use is met in an economy, i.e. how products and services reach consumers. There is an element of rethinking what are needs and what are actually wants, which may also fit into reduce
R2 reduce	Eco-efficiency in all its facets as well as energy reduction applied throughout the system. This strategy is an outcome for materials, energy use and emissions of applying the other r-strategies. It can extend from products through to buildings and other systems. Industrial symbiosis and other practices, e.g. cradle-to-cradle design also enable this outcome. Reduce is really about sufficiency and seems a difficult concept to grasp because it is about reducing the overall pile of resources used in an economy (and not just about stemming the inflow of new materials by recycling more for example)
R3 reuse	New business models and services promote affordable sharing of products and are a result of the refusal of the status quo, a rethink of how we consume and produce a reduction in material and energy use. Not only are products per se re-used by second markets but they are designed for this prolonged value (e.g. Sivaloganathan and Shahin 1999). Here we can also think about new roles of consumers, who play an active role in enabling reuse
R4 repair	Products designed for easy repairability by consumers or services, entailing of course not only a rethink of design, refusal of non-repairable products and a reduction in materials and energy, but also enabling the reuse of products. Design for repairability (Rosner and Ames 2014), modularity, repair cafes and business and employment based on this promoted. Includes right to repair and transparency about designs to enable repair. Disassembly as well as modularity and repairability
R5 refurbish	A used product has those elements replaced that enable it to continue in its original or upgraded function. Similar to R4 and R3 this entails a rethink of business models and design for refurbishment. This principle builds on existing examples and can be extended to include building refurbishment or retrofit
R6 remanufacture	Remanufacturing is a process in which components and products are sorted, selected, disassembled, cleaned, inspected and repaired or replaced before being reassembled and tested to function as good as new or better [217,218]. Arguably, remanufacturing has to follow a standardised industrial process that is "fully documented" and "capable of fulfilling the requirements established by the remanufacturer" (internationally agreed remanufacturing (Velenturf 2021). Design for remanufacture (e.g. Hatcher et al. 2011) is included here. This entails modularity regarding simplification in the diversity of product components for multi-purpose reassignment

(continued)

Table 1 (continued)

R7 repurpose	Products designed to enable their elements or totality can be used again in products with another function (e.g. Eike et al. 2020). Repurposing is part of a rethink of consumption and the economy and avoids recycling processes. It can be enabled by so-called digital marketplaces but also can entail the re-purposing of product elements, such as motors, into new applications
R8 recycle	Products are so designed with materials that they are fully recyclable into new products and markets. This entails that other r-strategies have been first deployed before the product or item passes to recycling. Note that the presence of recycling infrastructure and flows is not a guarantee that recycling is contributing to reduced emissions or consumption
R9 recover	Bioeconomy allows for products to re-enter the biosphere and incineration for energy recovery is avoided. Also, biogeochemical processes, which is about safely entrusting materials back to natural biogeochemical processes. An example could be a landfill where we're mining valuable materials and eventually only leaving materials that can safely return to become part of natural capital over longer periods of time

In sum, design in its expert and diffuse modes can contribute to promoting all aspects of the sustainable circular economy principles. For designers, this will entail multi-disciplinary collaborations and greater knowledge and experience of the social, environmental and economic theories and concepts listed above. While there will still be a place for traditional expert industrial design concerns, including eco-efficiency practice, material choices and life cycle thinking, new areas for design will include sufficiency thinking, business model design, systems thinking and participation in multi-stakeholder social innovation and policy. This will entail rethinking design education and the spaces of practice to include such complex environments. Such an approach will also entail inviting those outside of expert design to experience the value of participatory scenario-making for circular policy and futures and prototyping these social innovations. In this respect, circular design is consistent with existing proposals for a systems-oriented transition to a new economy, driven by new product service systems (EEA 2019).

Although only explicitly mentioned under principle 3, R-ladders are a set of ten principles that are traditionally defined in relation to product design. Within a circular society, they have a much broader application to all the principles. These R-strategies support each other and apply more broadly to all the spaces of the economy, environment and society listed above. As a result, R-strategies can be used as another way of describing principles that are consistent with sustainable circular economy (SCE) principles. Thus, circular design based on SCE principles leads to a new application of the R-ladder consistent with all the principles above.

While there are examples of circular design initiatives scattered in the literature and in training materials (Schmidt et al. 2020), they remain few and of limited scope. Bringing visualisation, prototyping and other design thinking skills to these environments while simultaneously expanding the design education remit to include all facets of sustainability transition will help achieve not only Papanek's but also Herbert Simon's agenda for the re-design of society.

Recent work on systems assessment of circular economy proposals and challenges puts these principles in perspective by identifying how the private, public and civic sectors can intervene in the current system through specific R-strategies that such sectors can employ (Bassi et al. 2021). These proposals, however, take a traditional view of R-strategy definitions unlike those proposed above, since they presume the continuation of a mainstream economy rather than new sufficiency-based consumption, and allow for high rates of recycling and waste inputs as well as energy recovery. There is also limited or no place for laws and regulations or exnovation as part of a strong intervention in the economy.

5 Manifesto for the New Profession: Circular Design

Based on interviews with practicing designers, Sumter et al. (2020) suggest "design for a circular economy can be seen as an independent, upcoming field in the ever-evolving sustainability domain, and for which specific competencies, tools, and methods are needed" (Sumter et al. 2020, p. 1561), and they argue for further work on what this might imply for higher education. While agreeing with this and other formulations of circular design (e.g. McAloone and Pigosso 2018), arguably the scope of this new field is far broader. Circular design as alluded to in the ten SCE principles and r-ladder, but particularly in principle 3, is a manifesto for a new design approach. Although there is a general awareness of the need for a new design profile to match the circular society goal written into the sustainable circular design agenda (Moreno 2016; Earley 2017), the specification of this knowledge and change remains largely limited to either revitalising design for sustainability or rather holistic accounts of socio-economic transformation through design.

Sustainable circular design works with multiple stakeholders to re-design industrial systems, supply chains, materials and products based on implementing the full list of R-strategies. Circular designers are aware of the positive and negative system-wide impacts of their actions. In collaboration with the private and public sectors, as well as civil society, they contribute to the phasing out of unsustainable practices, which itself is a product of government and business interventions. One approach has been to specify the multi-level government, business and society interventions and policies required to make circular design possible, as in the action plan for Scotland (Whicher et al. 2018). Thus, as noted, the expanded sense of circular r-strategies and the roles for design based on the ten sustainable design principles offer a more specific framework for the circular design than hitherto—a new understanding of the 'designing out waste' mantra of mainstream CE.

6 Conclusion

The development and implementation of a broad expert and diffuse design approach to circular economy are hinted at albeit in a scattered fashion in the literature. In account of circular design, R-strategies typically are limited to the material considerations of industrial and product design. The limited work on participatory and co-design approaches to policy and participation towards a circular society may acknowledge these professional industrial design concerns as a necessary but not sufficient approach to a sustainable circular economy. We suggest that recognising the broad remit of the practice of design, including some of its typical tools such as prototyping, visualisation and even business model design, may be a key way forward in redesigning the economy and society consistent with the aims of sustainable development. The r-strategies in this new approach constitute mindsets that can be brought to the task while the ten principles themselves articulate the system-wide changes in the economy, society and environment that need to be furthered. While various versions of new economic thinking, including the example of Doughnut Economics, post-Growth, and also Well-being economies and ideas, also propose disruptive and sometimes utopian visions of change, the approach we outline attempts to take a more pragmatic approach in acknowledging circularity as an important initiative but one which requires further work.

References

Andrews D (2015) The circular economy, design thinking and education for sustainability. Local Econ 30(3):305–315. https://doi.org/10.1177/0269094215578226

Asia-Pacific Economic Cooperation (2020) Circular economy roadmap—Malaysia's practice (2020/SOM1/PPSTI/029; Issue February). http://mddb.apec.org/Documents/2020/PPSTI/PPSTI1/20_ppsti1_029.pdf

Bassi AM, Bianchi M, Guzzetti M et al (2021) Improving the understanding of circular economy potential at territorial level using systems thinking. Sustain Prod Consum 27:128–140. https://doi.org/10.1016/j.spc.2020.10.028

Bauwens T, Hekkert M, Kirchherr J (2020) Circular futures: what will they look like? Ecol Econ 175:106703. https://doi.org/10.1016/j.ecolecon.2020.106703

Brown A, Haas W, Laubscher M et al (2018) The circularity Gap rgport: an analysis of the circular state of the global economy. Available at https://www.circularity-gap.world/. Accessed on 01 Nov 21

van den Bergh JCJM (2020) Six policy perspectives on the future of a semi-circular economy. Resour Conserv Recycl 160:104898. https://doi.org/10.1016/j.resconrec.2020.104898

Binet F, Coste-Manière I, Decombes C, Grasselli Y, Ouedermi D, Ramchandani M (2019) Fast fashion and sustainable consumption. In: Muthu SS (ed) Fast fashion, fashion brands and sustainable consumption. Springer Singapore, pp 19–35. https://doi.org/10.1007/978-981-13-1268-7_2

Birkeland J (2022) Nature positive: interrogating sustainable design frameworks for their potential to deliver eco-positive outcomes. Urban Sci 6:35. https://doi.org/10.3390/urbansci6020035

Biggs R, Schï M, Biggs D, et al (2012) Toward principles for enhancing the resilience of ecosystem services. Annu Rev Environ Resour 37:421–448. https://doi.org/10.1146/annurev-environ-051211-123836

Black R, Cullen K, Fay B et al (2021) Taking stock: a global assessment of net zero targets. Oxford, UK. Available at https://ec.europa.eu/clima/policies/strategies/2050_en. Accessed on 01 Nov 2023

Blum NU, Haupt M, Bening CR (2020) Why "Circular" doesn't always mean "Sustainable." Resour Conserv Recycl 162(June):105042. https://doi.org/10.1016/j.resconrec.2020.105042

Bocken NMP, de Pauw I, Bakker C, van der Grinten B (2016) Product design and business model strategies for a circular economy. J Ind Prod Eng 33(5):308–320. https://doi.org/10.1080/21681015.2016.1172124
Bocken NMP, Short SW (2020) Transforming business models: towards a sufficiency-based circular economy. In: Brandão M, Lazarevic D, Finnveden G (eds) Handbook of the circular economy. Edward Elgar Publishing, pp 250–265. https://doi.org/10.4337/9781788972727.00028
Boylston S (2019) Designing with society: a capabilities approach to design, systems thinking and social innovation. Taylor & Francis, London
Bressanelli G, Perona M, Saccani N (2019) Challenges in supply chain redesign for the circular economy: a literature review and a multiple case study. Int J Prod Res 57:7395–7422. https://doi.org/10.1080/00207543.2018.1542176
Britton G (2017) Co-design and social innovation: connections, tensions and opportunities (P. (Firm) issuing body (ed)), 1st ed. Routledge, London
Burkett I (2013) Using the business model canvas for social enterprise, 2nd edn. Griffith University Queensland. Available at https://www.griffith.edu.au/__data/assets/pdf_file/0042/996684/BMC-for-SE-2nd-Edition-Web.pdf. Accessed on 01 Nov 2023
Calisto Friant M, Vermeulen WJV, Salomone R (2020a) A typology of circular economy discourses: navigating the diverse visions of a contested paradigm. Resour Conserv Recycl 161(November 2019):104917. https://doi.org/10.1016/j.resconrec.2020.104917
Calisto Friant M, Vermeulen WJV, Salomone R (2020b) A typology of circular economy discourses: navigating the diverse visions of a contested paradigm. Resour Conserv Recycl 161(April):104917. https://doi.org/10.1016/j.resconrec.2020.104917
Camilleri MA (2018) Closing the loop for resource efficiency, sustainable consumption and production: a critical review of the circular economy. Int J Sustain Dev 21:1–17. https://doi.org/10.1504/IJSD.2018.100802
Calisto Friant M, Lakerveld D, Vermeulen WJV, Salomone R (2022) Transition to a sustainable circular plastics economy in the netherlands: discourse and policy analysis. Sustain 14:1–32. https://doi.org/10.3390/su14010190
Chapin FS, Carpenter SR, Kofinas GP et al (2010) Ecosystem stewardship: sustainability strategies for a rapidly changing planet. Trends Ecol Evol 25:241–249. https://doi.org/10.1016/j.tree.2009.10.008
Circular Economy Initiative Deutschland (2021) Circular-Economy-Roadmap-fuer-Deutschland
Corvellec H, Stowell AF, Johansson N (2021) Critiques of the circular economy. J Ind Ecol 1–12. https://doi.org/10.1111/jiec.13187
Corona B, Shen L, Reike D et al (2019) Towards sustainable development through the circular economy—A review and critical assessment on current circularity metrics. Resour Conserv Recycl 151:104498. https://doi.org/10.1016/j.resconrec.2019.104498
Crocker R (2018) From 'Spaceship earth' to the circular economy: the problem of consumption. In: Unmaking waste in production and consumption: towards the circular economy, 13–33. Emerald Publishing Limited. https://doi.org/10.1108/978-1-78714-619-820181003
van Dam K, Simeone L, Keskin D et al (2020) Circular economy in industrial design research: a review. Sustain 12:1–19. https://doi.org/10.3390/su122410279
Ddiba D, Andersson K, Koop SHA et al (2020) Governing the circular economy: assessing the capacity to implement resource-oriented sanitation and waste management systems in low- and middle-income countries. Earth Syst Gov 4. https://doi.org/10.1016/j.esg.2020.100063
D'Amato D, Droste N, Allen B, Kettunen M, Lähtinen K, Korhonen J, Leskinen P, Matthies BD, Toppinen A (2017) Green, circular, bio economy: a comparative analysis of sustainability avenues. J Clean Prod 168:716–734. https://doi.org/10.1016/j.jclepro.2017.09.053
Desing H, Brunner D, Takacs F et al (2020) A circular economy within the planetary boundaries: towards a resource—based, systemic approach. Resour Conserv Recycl 155:104673. https://doi.org/10.1016/j.resconrec.2019.104673
Eike R, Irick E, McKinney E et al (2020) Repurposing Design Process. In: Muthu S, Gardetti M (eds) Sustainability in the textile and apparel industries: production, processing, manufacturing & chemistry. Springer, Cham, pp 189–239

Huybrechts L, Benesch H, Geib J (2017) Institutioning: participatory design, co-design and the public realm. Co-Design 13:148–159. https://doi.org/10.1080/15710882.2017.1355006

den Hollander MC, Bakker CA, Hultink EJ (2017) Product design in a circular economy: development of a typology of key concepts and terms. J Ind Ecol 21(3):517–525. https://doi.org/10.1111/jiec.12610

Earley R (2017) Circular design futures. Des J 20:421–434. https://doi.org/10.1080/14606925.2017.1328164

Ellen MacArthur Foundation (2013) Towards the ciruclar economy: economic and business rationale for an accelerated transition, No 1(1). https://www.ellenmacarthurfoundation.org/assets/downloads/publications/Ellen-MacArthur-Foundation-Towards-the-Circular-Economy-vol.1.pdf

EMF (2015) Towards a circular economy: business rationale for an accelerated transition. Ellen MacArthur Foundation. Available at https://www.ellenmacarthurfoundation.org/towards-the-circular-economy-vol-1-an-economic-and-business-rationale-for-an. Accesed on 01 Nov 2023

European Environmental Agency (EEA) (2019) Sustainability transitions: policy and practice. https://doi.org/10.2800/641030

Fan Y Van, Lee CT, Lim JS et al (2019) Cross—disciplinary approaches towards smart, resilient and sustainable circular economy. J Clean Prod 232:1482–1491. https://doi.org/10.1016/j.jclepro.2019.05.266

Fitch-Roy O, Benson D, Monciardini D (2019) Going around in circles? Conceptual recycling, patching and policy layering in the EU circular economy package. Env Polit 00:1–21. https://doi.org/10.1080/09644016.2019.1673996

Folke C, Carpenter SR, Walker B, Scheffer M, Chapin T, Rockstrom J (2010) Resilience thinking: integrating resilience, adaptability and transformability. Ecol Soc 15(4). art20

Fossati EC, Sureau S, Pel B et al (2022) Exnovation: imagining sustainable transitions differently in Brussels. Brussels Stud 0–20. https://doi.org/10.4000/brussels.6327

Fry T (2009) Design futuring: sustainability, ethics and new practice. Bloomsbury Academic Publishing, London

Geels FW (2011) The multi-level perspective on sustainability transitions: responses to seven criticisms. Environ Innov Soc Trans 1(1):24–40. https://doi.org/10.1016/j.eist.2011.02.002

Geissdoerfer M, Savaget P, Bocken NMPP et al (2017) The circular economy—a new sustainability paradigm? J Clean Prod 143:757–768. https://doi.org/10.1016/j.jclepro.2016.12.048

Halstenberg FA, Stark R (2019) Introducing product service system architectures for realizing circular economy. Procedia Manuf 33:663–670. https://doi.org/10.1016/j.promfg.2019.04.083

Hales D (2013) Design fictions an introduction and provisional taxonomy. Digit Creat 24:1–10. https://doi.org/10.1080/14626268.2013.769453

Hatcher GD, Ijomah WL, Windmill JFC (2011) Design for remanufacture: a literature review and future research needs. J Clean Prod 19:2004–2014. https://doi.org/10.1016/j.jclepro.2011.06.019

Harris S, Martin M, Diener D (2021) Circularity for circularity's sake? Scoping review of assessment methods for environmental performance in the circular economy. Sustain Prod Consum 26:172–186. https://doi.org/10.1016/j.spc.2020.09.018

Hardy C, Thomas R (2015) Discourse in a material world. J Manag Stud 52:680–696. https://doi.org/10.1111/joms.12113

Haas W, Krausmann F, Wiedenhofer D, Lauk C, Mayer A (2020) Spaceship earth's odyssey to a circular economy—a century long perspective. Resour Conserv Recycl 163:105076. https://doi.org/10.1016/j.resconrec.2020.105076

Haucke FV (2018) Smartphone-enabled social change: evidence from the Fairphone case? J Clean Prod 197:1719–1730. https://doi.org/10.1016/j.jclepro.2017.07.014

Hobson K (2021) The limits of the loops: critical environmental politics and the Circular Economy. Env Polit 30:161–179. https://doi.org/10.1080/09644016.2020.1816052

Hobday M, Boddington A, Grantham A (2012) Policies for design and policies for innovation: contrasting perspectives and remaining challenges. Technovation 32:272–281. https://doi.org/10.1016/j.technovation.2011.12.002

Holzinger H (2020) More efficiency is not enough. Capabilities and limits of the circular economy. In: the circular economy in the european union. Springer International Publishing, Cham, pp 187–206

Horvath B, Bahna M, Fogarassy C (2019) The ecological criteria of circular growth and the rebound risk of closed loops. Sustainability 11:2961. https://doi.org/10.3390/su11102961

Homrich AS, Galvão G, Abadia LG, Carvalho MM (2018) The circular economy umbrella: trends and gaps on integrating pathways. J Clean Prod 175:525–543. https://doi.org/10.1016/j.jclepro.2017.11.064

Howlett M (2020) Challenges in applying design thinking to public policy: dealing with the varieties of policy formulation and their vicissitudes. Policy Polit 48(1):49–65. https://doi.org/10.1332/030557319X15613699681219

Jaeger-Erben M, Jensen C, Hofmann F, Zwiers J (2021) There is no sustainable circular economy without a circular society. Resour Conserv Recycl 168(January):105476. https://doi.org/10.1016/j.resconrec.2021.105476

Jackson T (2009) Prosperity without growth? The transition to a sustainable economy. Routledge. London, UK

Kahane A (2012) Transformative scenario planning: changing the future by exploring alternatives. Strateg Leadersh 40:19–23. https://doi.org/10.1108/10878571211257140

Kanda W, Kuisma M, Kivimaa P, Hjelm O (2020) Conceptualising the systemic activities of intermediaries in sustainability transitions. Environ Innov Soc Transitions 36:449–465. https://doi.org/10.1016/j.eist.2020.01.002

Kayikci Y, Kazancoglu Y, Lafci C, Gozacan N (2021) Exploring barriers to smart and sustainable circular economy: the case of an automotive eco-cluster. J Clean Prod 314:127920. https://doi.org/10.1016/j.jclepro.2021.127920

Kirchherr J, Reike D, Hekkert M (2017) Conceptualizing the circular economy: an analysis of 114 definitions. Resour Conserv Recycl 127:221–232. https://doi.org/10.1016/j.resconrec.2017.09.005

Kishita Y, Hara K, Uwasu M, Umeda Y (2016) Research needs and challenges faced in supporting scenario design in sustainability science: a literature review. Sustain Sci 11:331–347. https://doi.org/10.1007/s11625-015-0340-6

Kivimaa P, Boon W, Hyysalo S, Klerkx L (2019) Towards a typology of intermediaries in sustainability transitions: a systematic review and a research agenda. Res Policy 48:1062–1075. https://doi.org/10.1016/j.respol.2018.10.006

Komiyama H, Takeuchi K (2006) Sustainability science: building a new discipline. Sustain Sci 1(1):1–6. https://doi.org/10.1007/s11625-006-0007-4

Kothari A, Demaria F, Acosta A (2014) Buen vivir, degrowth and ecological swaraj: alternatives to sustainable development and the green economy. Development 57:362–375. https://doi.org/10.1057/dev.2015.24

Korhonen J, Nuur C, Feldmann A, Birkie SE (2018) Circular economy as an essentially contested concept. J Clean Prod 175:544–552. https://doi.org/10.1016/j.jclepro.2017.12.111

Koutra C (2010) Rapid situation analysis: a hybrid, multi-methods, qualitative, participatory approach to researching tourism development phenomena. J Sustain Tour 18(8):1015–1033. https://doi.org/10.1080/09669582.2010.497221

Kumar A, Lodha D, Mahalingam A, et al (2016) Using 'design thinking' to enhance urban redevelopment: a case study from India. Eng Proj Organ J 3727:1–11. https://doi.org/10.1080/21573727.2016.1155445

Lamberton G (2005) Sustainable sufficiency—An internally consistent version of sustainability. Sustain Dev 13:53–68. https://doi.org/10.1002/sd.245

Lacovidou E, Hahladakis JN, Purnell P (2021) A systems thinking approach to understanding the challenges of achieving the circular economy. Environ Sci Pollut Res 28:24785–24806

Leipold S, Weldner K, Hohl M (2021) Do we need a 'circular society'? Competing narratives of the circular economy in the French food sector. Ecol Econ 187:107086. https://doi.org/10.1016/j.ecolecon.2021.107086

Leipold S (2021) Transforming ecological modernization 'from within' or perpetuating it? The circular economy as EU environmental policy narrative. Env Polit 30:1045–1067. https://doi.org/10.1080/09644016.2020.1868863
Ljungberg L (2007) Materials selection and design for development of sustainable products. Mater Des 28:466–479. https://doi.org/10.1016/j.matdes.2005.09.006
Lewandowski M (2016) Designing the business models for circular economy-towards the conceptual framework. Sustain 8:1–28.
Lofthouse V, Prendeville S (2018) Human-Centred design of products and services for the circular economy–a review. Des J 21(4):451–476. https://doi.org/10.1080/14606925.2018.1468169
Loorbach D, Frantzeskaki N, Avelino F (2017) Sustainability transitions research: transforming science and practice for societal change. Annu Rev Environ Resour 42:599–626. https://doi.org/10.1146/annurev-environ-102014-021340
Makov T, Vivanco DF (2018) Does the circular economy grow the pie? The case of rebound effects from smartphone reuse. Front Energy Res 6(May):1–11. https://doi.org/10.3389/fenrg.2018.00039
Manzini E (2015) Design when everybody designs: an introduction to design for social innovation. MIT Press.
McDowall W, Geng Y, Huang B, Barteková E, Bleischwitz R, Türkeli S, Kemp R, Doménech T (2017) Circular economy policies in China and Europe. J Ind Ecol 21(3):651–661. https://doi.org/10.1111/jiec.12597
McGann M, Blomkamp E, Lewis JM (2018) The rise of public sector innovation labs: experiments in design thinking for policy. Policy Sci 51:249–267. https://doi.org/10.1007/s11077-018-9315-7
Melles G (2021) Figuring the transition from circular economy to circular society in Australia. Sustainability (Switzerland) 13(19). https://doi.org/10.3390/su131910601
Meadows DH (2008) Thinking in systems: a primer. Chelsea Green Pulbishing, White River Junction, NJ
Melles G, Vere ID, Misic V (2011) Socially responsible design: thinking beyond the triple bottom line to socially responsive and sustainable product design. CoDesign 7(3–4):37–41. https://doi.org/10.1080/15710882.2011.630473
Melles G, Wölfel C, Krzywinski J, Opeskin L (2022) Expert and diffuse design of a sustainable circular economy in two German circular roadmap projects. Soc Sci 11(9). https://doi.org/10.3390/socsci11090408
Mishra S, Singh SP, Johansen J, Cheng Y, Farooq S (2019) Evaluating indicators for international manufacturing network under circular economy. Manag Decis 57(4):811–839. https://doi.org/10.1108/MD-05-2018-0565
Moreno M, De los Rios C, Rowe Z, Charnley F (2016) A conceptual framework for circular design. Sustainability (Switzerland) 8:937. https://doi.org/10.3390/su8090937
Mononen L (2017) Understanding future designer thinking Systems thinking and its contribution to understanding future designer thinking. Des J 20:S4529–S4538. https://doi.org/10.1080/14606925.2017.1352949
Mulgan G (2010) Measuring social value. Standford Soc Innov Rev Summer 41–55. https://doi.org/10.4324/9780203167380-10
Papanek V (1971) Design for the real world: human ecology and social change, 2nd ed. Bantam Books. http://www.amazon.com/Design-Real-World-Ecology-Social/dp/0897331532
Platform for Accelerating the Circular Economy (PACE) (2020) The circularity gap report 2020: when circularity goes from bad to worse: the power of countries to change the game. https://www.circularity-gap.world/
Price Waterhouse Cooper (2019) The road to circularity: why a circular economy is becoming the new normal. Available at https://www.pwc.nl/en/assets/documents/pwc-the-road-to-circularity-en.pdf. Accessed on 01 Nov 2023
Rauschmayer F, Bauler T, Schäpke N (2015) Towards a thick understanding of sustainability transitions—Linking transition management, capabilities and social practices. Ecol Econ 109:211–221. https://doi.org/10.1016/j.ecolecon.2014.11.018

Ralph Boch, Gallen J, Hempel N (2020) Wege zu einer circular society: potenziale des social design für gesellschaftliche transformation. https://socialdesign.de/wp-content/uploads/2020/04/200420_HSS_Paper_CircularSociety_online.pdf
Raworth K (2017) Doughnut economics: seven ways to think like a 21st century economics. Random House Business Books
Reike D, Vermeulen WJV, Witjes S (2018) The circular economy: new or refurbished as CE 3.0?—exploring controversies in the conceptualization of the circular economy through a focus on history and resource value retention options. Resour Conserv Recycl 135(August 2017):246–264. https://doi.org/10.1016/j.resconrec.2017.08.027
Rockström J, Steffen W (2009) Planetary boundaries: exploring the safe operating space for humanity. Ecol Soc 14(2):32. http://www.ecologyandsociety.org/vol14/iss2/art32/main.html
Saidani M, Yannou B, Leroy Y, Cluzel F (2017) How to assess product performance in the circular economy? Proposed requirements for the design of a circularity measurement framework. Recycling 2(1). https://doi.org/10.3390/recycling2010006
Sands G (2015) Measuring the prosperity of cities. Habitat Int 45:1–2. https://doi.org/10.1016/j.habitatint.2014.06.016
Sanders EB-N, Stappers PJ (2008) Co-creation and the new landscapes of design. CoDesign 4(1):5–18. https://doi.org/10.1080/15710880701875068
Sanoff H (2005) Community participation in riverfront development. CoDesign 1:61–78. https://doi.org/10.1080/15710880512331326022
Saavedra YMB, Iritani DR, Pavan ALR, Ometto AR (2018) Theoretical contribution of industrial ecology to circular economy. J Clean Prod 170:1514–1522. https://doi.org/10.1016/j.jclepro.2017.09.260
Schandl H, King S, Walton A, Kaksonen A, Tapsuwan S, Baynes T (2021) Circular economy roadmap for plastics, glass, paper and tyres–pathways for unlocking future growth opportunities for Australia (Issue January). CSIRO Publishing
Schmidt K, Bundgaard A, Hirsbak S et al (2020) Product-Service development for circular economy and sustainability course. LNEG—Laboratório Nacional de Energia e Geologia, I.P. Portugal
Schröder P, Bengtsson M, Cohen M, Dewick P, Hoffstetter J, Sarkis J (2019) Degrowth within—aligning circular economy and strong sustainability narratives. Resour Conserv Recycl 146(April):190–191. https://doi.org/10.1016/j.resconrec.2019.03.038
Schroeder P, Anggraeni K, Weber U (2019) The relevance of circular economy practices to the sustainable development goals. J Ind Ecol 23:77–95. https://doi.org/10.1111/jiec.12732
Seddon N, Smith A, Smith P et al (2021) Getting the message right on nature-based solutions to climate change. Glob Chang Biol 27:1518–1546. https://doi.org/10.1111/gcb.15513
Selloni D, Corubolo M (2017) Design for social enterprises: how design thinking can support Social Innovation within Social Enterprises. Des J 20:775–794. https://doi.org/10.1080/14606925.2017.1372931
Shaw A, Sheppard S, Burch S et al (2009) Making local futures tangible—Synthesizing, downscaling, and visualizing climate change scenarios for participatory capacity building. Glob Environ Chang 19:447–463. https://doi.org/10.1016/j.gloenvcha.2009.04.002
Smith BG (2008) Developing sustainable food supply chains. Philos Trans R Soc Lond B Biol Sci 363:849–861. https://doi.org/10.1098/rstb.2007.2187
Stahel WR (2010) The performance economy. In: The performance economy, 2nd ed. Palgrave Macmillan UK. https://doi.org/10.1057/9780230288843
Stahel WR (2020) History of the circular economy. The historic development of circularity and the circular economy. In: The circular economy in the European union. Springer International Publishing, pp 7–19. https://doi.org/10.1007/978-3-030-50239-3_2
Sterman JD (2000) Business dynamics: systems thinking and modeling for a complex world. Irwin McGraw-Hill, Boston, MA
Sumter D, Bakker C, Balkenende R (2018) The role of product design in creating circular business models: a case study on the lease and refurbishment of baby strollers. Sustain 10. https://doi.org/10.3390/su10072415

Sumter D, de Koning J, Bakker C, Balkenende R (2020) Circular economy competencies for design. Sustainability (switzerland) 12(4):1–16. https://doi.org/10.3390/su12041561
Temesgen A, Storsletten V, Jakobsen O (2019) Circular economy—reducing symptoms or radical change? Philos Manag. https://doi.org/10.1007/s40926-019-00112-1
Teasdale S (2012) What's in a name? Making sense of social enterprise discourses. Public Policy Adm 27:99–119. https://doi.org/10.1177/0952076711401466
UNEP (2003) Life cycle assessment. Available at papers2://publication/uuid/C59E569D-DC25-43FA-BFB1-27A716FDB9B1. Accessed on 01 Nov 2023
van der Velden M (2021) 'Fixing the world one thing at a time': community repair and a sustainable circular economy. J Clean Prod 304:127151. https://doi.org/10.1016/j.jclepro.2021.127151
Velenturf APM, Jopson JS (2019) Making the business case for resource recovery. Sci Total Environ 648:1031–1041. https://doi.org/10.1016/j.scitotenv.2018.08.224
Velenturf APM (2021) A framework and baseline for the integration of a sustainable circular economy in offshore wind. Energies 14(17). https://doi.org/10.3390/en14175540
Velenturf APM, Archer SA, Gomes HI, Christgen B, Lag-Brotons AJ, Purnell P (2019) Circular economy and the matter of integrated resources. Sci Total Environ 689:963–969. https://doi.org/10.1016/j.scitotenv.2019.06.449
Vezzoli C, Kohtala C, Srinivasan A, et al (2017) Product-service system design for sustainability. In: Carlo Vezzoli, Kohtala C, Srinivasan A et al (eds) Product-Service System Design for Sustainability, 1st edn. Routledge, pp 49–86
Vonk L (2018) Paying attention to waste: apple's circular economy. Continuum (NY) 32:745–757. https://doi.org/10.1080/10304312.2018.1525923
Velenturf APM, Purnell P (2021) Principles for a sustainable circular economy. Sustain Prod Consum 27:1437–1457. https://doi.org/10.1016/j.spc.2021.02.018
Wanner T (2015) The new 'passive revolution' of the green economy and growth discourse: maintaining the 'sustainable development' of neoliberal capitalism. New Polit Econ 20:21–41. https://doi.org/10.1080/13563467.2013.866081
Ward JD, Sutton PC, Werner AD, Costanza R, Mohr SH, Simmons CT (2016) Is decoupling GDP growth from environmental impact possible? PLoS ONE 11:1–14. https://doi.org/10.1371/journal.pone.0164733.
Wastling T, Charnley F, Moreno M (2018) Design for circular behaviour: considering users in a circular economy. Sustainability 10. https://doi.org/10.3390/su10061743
World Economic Forum (2020) Raising Ambitions: a new roadmpa for the automotive circular economy (Issue December). https://www.accenture.com/_acnmedia/PDF-146/Accenture-and-WEF-Raising-Ambitions-PoV.pdf#zoom=50
Winans K, Kendall A, Deng H (2017) The history and current applications of the circular economy. Concept Renew Sustain Ener Rev 68:825–833. https://doi.org/10.1016/j.rser.2016.09.123.
Wiedmann TO, Schandl H, Lenzen M et al (2015) The material footprint of nations. Proc Natl Acad Sci USA 112:6271–6276. https://doi.org/10.1073/pnas.1220362110
Whicher A, Harris C, Beverley K, Swiatek P (2018) Design for circular economy: developing an action plan for Scotland. J Clean Prod 172:3237–3248. https://doi.org/10.1016/j.jclepro.2017.11.009
Wu KJ, Hou W, Wang Q et al (2022) Assessing city's performance-resource improvement in China: a sustainable circular economy framework approach. Environ Impact Assess Rev 96:106833. https://doi.org/10.1016/j.eiar.2022.106833

Practice Perspective Implications for Sustainable Circular Economy Transitions

8

Olamide Shittu and Christian Nygaard

1 Introduction

Many cities have attempted to advance sustainability transitions by implementing policies that shape market forces and influence individual behaviours (Davies and Doyle 2015). However, while necessary, studies have shown that these efforts have been insufficient to incentivise the transformation of ways of living and practices and significant systemic change, for instance, in waste infrastructure (Marres 2011). Moreover, households still face systemic barriers to translating sustainability education into real choices and actions, changing actual practices (Lavelle, Rau and Fahy 2015). For instance, while low-income households maintain and reuse plastic materials due to financial considerations, the constraints often imposed by inadequate access to spatiotemporal (such as storage) and socio-economic (such as knowledge) resources prevent them from performing other activities, such as sorting and repair that otherwise might form part of a sustainable CE practice (Shittu 2023a; Shittu et al. 2021).

Practice perspectives move beyond the focus on structures or individual actions in sustainability transitions and instead focus on the bundle of activities and sociomaterial arrangements that shape social reality (Strengers and Maller 2014). Here, sociomaterial arrangements refer to the "relationships between human [activities] and material arrangements [set-ups or configurations], and how these relationships emerge in a practice" (Nyström et al. 2016, p. 3). Such an approach recognises

O. Shittu (✉)
Circular Economy Research Fellow, Centre for Urban Transitions, Swinburne University of Technology, PO Box 218, Hawthorn, VIC 3122, Australia
e-mail: oshittu@swin.edu.au

C. Nygaard
Centre for Urban Transitions, Swinburne University of Technology, Melbourne, VIC, Australia

© The Author(s), under exclusive license to Springer Nature Singapore Pte Ltd. 2024
G. B. Melles and C. Wölfel (eds.), *Design for a Sustainable Circular Economy*, Design Science and Innovation, https://doi.org/10.1007/978-981-99-7532-7_8

the interaction between materials (and materiality of things), meanings, and competencies in combining disparate daily activities in a sustainable CE practice. Specifically, practice theory focuses on the role of materiality in shaping patterns of production, consumption, and disposal, which is often missing in transitions and behaviour change literature.

The doctoral research (Shittu 2022) summarised in this chapter focuses on household plastic consumption. It explores how one might conceptualise sustainable CE practices and its implication for transitioning cities to sustainable CE platforms. Bringing together conceptual and empirical, this chapter argues that recognising the sociomaterial and spatiotemporal context of practices in individual cities is critical before designing sustainable CE policies or strategies. Sustainability solutions are, thus, substantively local due to the entrenchment of sociomaterial and spatiotemporal arrangements in existing social structures. The implication is that an uncritical adoption of CE policies and approaches between cities and the global north and south is likely unproductive and ineffectual.

This chapter examines two critical insights from a practice perspective on sustainable CE analysis in research, policy, and practice. First, sustainable CE practices do not easily conform to binary scales such as 'sustainable' and 'unsustainable'. As a result, we cannot robustly identify the bundles(s) of daily activities that might constitute a sustainable CE practice. Hence, the chapter proposes an alternate analogy of sustainable CE practices, which exists on three different variations and stages of performance. These are input assemblage, input combinations, and outputs and outcomes. Second, transitioning to sustainable CE calls for innovation by combination and resourcefulness in socio-technical systems. This entails going beyond business-as-usual to re-imagine new, radical, and disruptive practice configurations in socio-technical systems, including markets, product and service design, technological innovations, infrastructure provision, and community services and policy landscapes.

The remainder of the chapter is structured as follows: Sect. 2 presents a brief background on practice theory; Sect. 3 discusses insights from Shittu (2022) on sustainable CE; Sect. 4 examines sustainability considerations for nuanced analyses of practices; Sect. 5 discusses the importance of innovation by combination to achieve sustainable CE transitions, and Sect. 6 concludes the chapter.

2 Practice Theory

Although not a unified body of theory, practice theory neither focuses on structures nor individual actions but on the range of activities that order and give meaning to social living. It can, therefore, be considered a middle-range perspective (Kuijer 2014). A popular definition of a practice is provided by Reckwitz (2002, p. 249) thus:

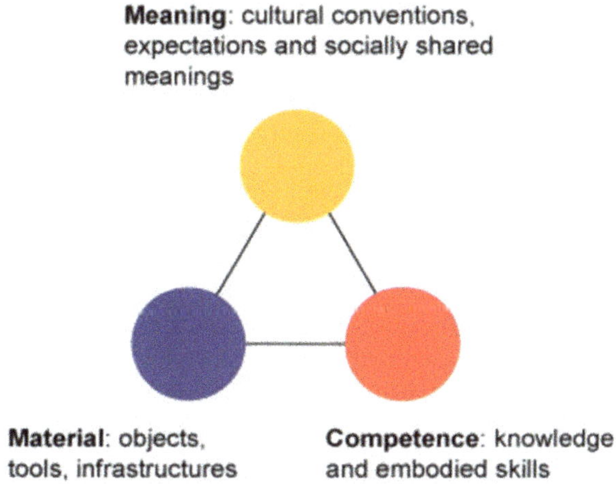

Fig. 1 Elements of practices. *Source* Shove et al. (2012)

> A routinized type of behaviour which consists of several elements, interconnected to one other: forms of bodily activities, forms of mental activities, 'things' and their use, a background knowledge in the form of understanding, know-how, states of emotion and motivational knowledge.

As shown in Fig. 1, Shove et al. (2012) describe a practice as composed of essential elements: material, meaning, and competence. Material varies from small objects like household plastic items to significant technologies and infrastructure such as the electricity grid of a city. The human body is also categorised as a material element of practice. Meanings encompass affective values, expectations, rules, and socio-cultural conventions that attach subjective significance to practices. In this respect, CE and sustainability can separately and jointly constitute meanings. Meanwhile, competence refers to the knowledge and embodied ability to perform a practice proficiently.

To illustrate, the performance of drinking as a practice could involve a cup, liquid, and the human body as materials; the meanings of sustenance, pleasure, social bonding, and healthy living; and knowledge and embodied skills such as ensuring the right quality, quantity, time, context, and the practicalities of transferring water into the mouth.

Critically, practice theory attempts to decentralise humans in social theorisation by regarding them as practitioners rather than actors. However, to Shove et al. (2007), practices recruit individuals as practitioners, or carriers, through their family, profession, social group, and cultural background. Moreover, bundles of practices form and shape the texture of daily life by manifesting as instances of consumption or activity. Therefore, examining bundles of practices that individuals or groups engage in is critical to understanding and potentially shaping why, how, and when individuals might perform a sustainable CE practice.

Practices are believed to manifest in two interacting forms—practices-as-entity and practices-as-performance (Schatzki 1996; Shove et al. 2007; Warde 2005). The former is a broader understanding of practice as a form that exists and endures in space and time with provisions of guidelines as to how they should be enacted in performance (Shove et al. 2007). This might include culturally or socially anchored meanings in food preparation, living spaces or accumulation of material goods in signalling social status. On the other hand, practices-as-performance is the enactment of practice through the unique process of combining the abovementioned elements in a socio-spatial and temporal context (Warde 2005). These include the functional (goal attainment) side of practices. Understanding sustainable CE through a practice perspective thus requires focusing on elements that constitute daily practices on the one hand (practices-as-performance) and institutional complexes on the other (practices-as-entity).

Furthermore, sustainable CE transitions require a detailed connection between seemingly mundane practices-as-performance and abstract practices-as-entity. Jaeger-Erben and Offenberger (2014) applied the 'iceberg metaphor' to achieve this. This framework describes the visible aspect of daily consumption as the tip of the iceberg, which may be an instance of the performance of a practice (Shove et al. 2012). However, it is intricately "embedded in socio-cultural and socio-technical settings" (Jaeger-Erben and Offenberger 2014, p. 166).

Within sustainability transitions, Jaeger-Erben and Offenberger's study (2014) reveals that changes in everyday consumption usually accompany adopting new sustainability practices in households, which often precede a life event. Life events activate tangible and intangible social fields with specific spatial, cultural, and historical evolution and comprise practices-as-entity (Jaeger-Erben and Offenberger 2014). Furthermore, adopting new sustainability practices also results in changes to the justification for and meaning of consumption, especially as it becomes embedded in the web of other practices within the household. This, therefore, emphasises the importance of context as the sustainable CE practice may manifest in diverse ways within different spatiotemporal and sociomaterial arrangements.

While practice theory is increasingly applied to sustainability transitions, practice theory is also criticised for focusing too much on systemic complexities and divergences retrospectively and too little on "direct and practice-oriented impact for change" (Öztekin and Gaziulusoy 2020). Most empirical studies adopting practice theory often approach understanding change by studying past or ongoing events. However, the methodology to operationalise practices for creating future impacts or direct interventions is still emerging. Direct impact for change requires solutions to reorganise the internal processes of a practice, substitute unsustainable practices with sustainable ones and reconfigure bundles of practices in new ways (Öztekin and Gaziulusoy 2020: 206). Reconfiguring practices involves transforming "user motivations, reasons, needs and wants" that engenders the entire network of practice bundles, which could then "generate fundamental shifts in wholes of practices and everyday lifestyles, and extensively contribute to sustainability transitions" (Öztekin and Gaziulusoy 2020: 206).

The following two sections summarise critical insights from doctoral research on the practice theory perspective and the transition of urban households to sustainable CE for plastics.

3 Transforming Plastic Consumption in Households

The emergence and presence of sustainability concerns in urban households, including household sustainability knowledge and attitudes, do not necessarily result in sustainable lifestyles. Specifically, households cannot consume sustainably, not necessarily because they lack sustainability knowledge, but due to the complex interconnection of socio-cultural practices (Sole and Wagner 2018; Paddock 2017). Therefore, policies targeting individual characteristics, or market-based approaches to systemic change, have largely not resulted in sustainable consumption.

Concomitantly, consumption studies adopting theories of practice tend to focus on non-material aspects such as energy-related practices. Nicolini (2009) calls for practice research to zoom in on material arrangements in social domains. Moreover, consumption studies literature could not identify any substantial difference in the unsustainable environmental impact of material consumption between low-income and high-income households (Buhl et al. 2018; Cai, Liu and Zhang 2019). The additional socio-economic and environmental difficulties they face in adopting sustainable lifestyles make transitioning to sustainable consumption more challenging for the former.

Shittu (2022) interrogates plastic-related practices in urban low-income households and draws insights for grassroots strategies for sustainable CE transitions in cities to advance understanding of materiality in practice theory. Beyond dominant Western case studies, data from case studies of low-income households in an emerging economy presents new contexts and insights into how a ubiquitous material like plastic could materialise and routinise household practices.

Empirical findings show that plastic materiality facilitates the performance of household practices through their corporeal, spatiotemporal, and functional dimensions. As actants in household practice performance (Strengers, Nicholls, and Maller 2016), plastic materials anchor meanings and skills through their unique corporal features. This functionality makes plastic indispensable to households and problematic for implementing sustainable CE interventions.

Therefore, achieving sustainable CE for plastics requires transforming plastic-related practices and integrating them into broader sustainability transitions of cities. In this sense, plastic-related sustainable CE practices include (Fig. 2): protractive practices, which extend the life cycle of plastic; contractive practices, which reduce the amount of plastic in circulation; and regenerative practices, which transform plastic waste into new products. As previously mentioned, while low-income households perform some protractive activities such as maintenance and reuse, they are influenced mainly by socio-economic factors rather than environmental considerations. Also, low-income households' sociomaterial arrangements inhibit their contractive and regenerative activities, including sharing and recycling. This chapter extends this conceptual framework to include all technical and material processes in sustainable CE.

Aside from providing waste infrastructures, addressing sustainable CE practice issues involves adopting grassroots environmental governance strategies for

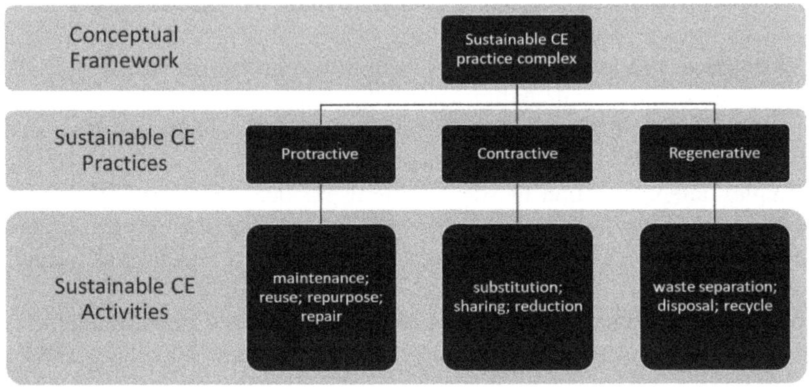

Fig. 2 Elements of sustainable CE practices. *Source* Shittu (2022)

urban transitions. These include addressing environmental justice issues in urban households and leveraging communities of practice (Wenger-Trayner and Wenger-Trayner 2015). Addressing environmental justice issues includes improving access to sustainability information, public participation in sustainability consultations, and providing sustainability infrastructure. Meanwhile, communities of practice can promote sustainable circular transitions by scaling up circular practices, sustaining circular practice complexes, and directing the transformation of circular practices.

4 A Conceptual Framing of Sustainable CE Practices: Three Stages of Performance

Analysis of sustainable CE practices (or behaviours in traditional theoretical approaches) does not easily conform to binary scales of 'sustainable' and 'unsustainable' (see Pocol et al. 2020; Lee et al. 2016; Zorell 2020). This is because a simple division of CE practices into binary sustainability scales ignores the ongoing and context-specific processes that occur during the interaction of elements of practices (materials, skills, and meanings) that result in environmental outcomes. Instead, sustainable CE practices can be conceptualised as existing in three different variations and stages of performance: input assemblage, input combinations, and outputs and outcomes (see Fig. 3). This conceptualisation can apply to the performance of sustainable CE practices in any societal domain, including households, businesses, and politics.

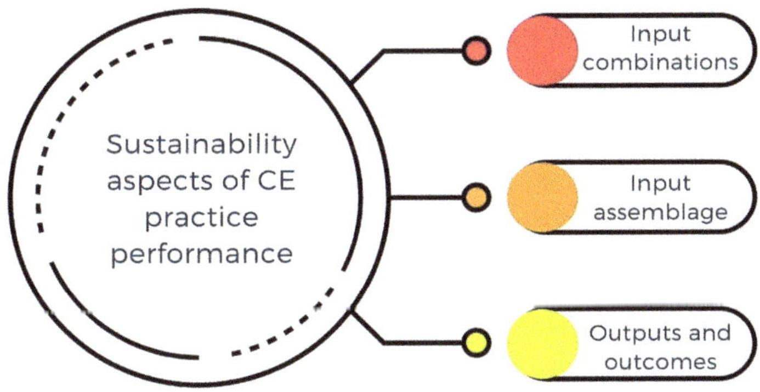

Fig. 3 Sustainability considerations in stages of CE practice performance. *Source* Authors' design

4.1 Input Assemblage in Sustainable CE Practice Performance

Practitioners may obtain or assemble input elements sustainably to perform a sustainable CE practice. This can be the use of sustainable materials, the sustainable procurement of materials, the acquisition of sustainability skills, and the setting of sustainability aspirations. Nonetheless, not all sustainable CE practices necessarily require sustainable inputs. For instance, the performance of recycling as a sustainable CE practice may involve a range of materials not classified as 'sustainable' and with, a priori, significant environmental footprints, such as plastics. In such cases, sustainable CE practices emphasise the transformation processes and a sustainable outcome.

Notably, the input of a practice performance is often the outcome or by-product of another ongoing or previously completed (set of) practice(s). For households, gathering the practice elements for cooking, for instance, could be preceded by the performance of shopping practices or participation in urban agriculture, the embodiment of cooking skills through professional or informal training practices, and the adoption of meanings of sustenance and cultural culinary art through socialisation. Likewise, acquiring input materials for a business can involve several complex practices such as design, extraction, business-to-business purchase, transportation, communication, and negotiation.

Consequently, a sustainable CE initiative can become ineffective if the inputs are supplied from unsustainable, close-knitted practices. Recycling systems in cities are a good illustration of this. Recycling systems in cities suffer from the lack of markets for recycled products and preceding factors such as the improper sorting of household and industrial waste (Ibrahim 2020; Yoada et al. 2014). This results in the contamination of potential recyclables, which puts significant pressure on the recycling system, and most often, plastic waste pollutes the environment.

As a component of regenerative practices, sorting is one of the sustainable CE activities that must be embedded in household practices to transfer clean input to the recycling system (Shittu et al. 2021). However, where the skills and motivation to carry out regenerative practices are present, many lower-income households face difficulty performing these CE practices due to their lack of access to adequate storage spaces. Furthermore, in many global south contexts, access to materials (such as hot water) might further inhibit the performance of recycling practices. Beyond that, the performance of other plastic-related sustainable CE practices, such as protractive and contractive practices in households and industries, will reduce the pressure on the recycling system and the energy footprint of material regeneration processes. Thus, researchers, policymakers, and other sustainability stakeholders must consider the extent to which a CE practice is integrated with others when designing intervention programmes.

4.2 Input Combination in Sustainable CE Practice Performance

Combining the input elements can also reveal some sustainability aspects of a CE practice. Here, the materials are sustainably and skilfully commingled following sustainability principles to achieve a desired goal, such as reducing the negative environmental impact of practices. For households and businesses, this could be in terms of using the correct quantity of materials, avoiding material damage, maximising material functionalities, proper channelling of unwanted materials, and other activities that prevent environmental leakage and improve resource efficiency.

For instance, many low-income households exercise caution in handling plastic materials when performing daily practices to extend their lifecycles (Shittu 2023a; Shittu et al. 2021). Specific examples from low-income households include avoiding using a plastic spoon to stir food when cooking and filling a plastic bucket with the water quantity that corresponds with its capacity. Thus, from a practice perspective, the sustainable combination of materials during the performance of a CE practice is heavily reliant on the level of embodied sustainability competences and the cohesion of sustainability meanings and values.

The detailed sustainability evaluation of input combinations of CE practices is essential for several practical reasons. As also argued by Buhl et al. (2018), the resource intensification of consumption and, consequently, the material and waste footprints of households are less about the income of households and more about the input combinations and performance processes of individual household practices. Moreover, even with sustainable technologies designed to reduce households' energy footprint, the lack of adequate knowledge and skills in using these technologies while performing energy-related practices inhibit their effectiveness (Herrmann et al. 2018; Shittu 2020, 2023a). The focus and emergence of sustainable materials and technologies, while a critical element of a shift in practices, remains incomplete without an accompanying embodiment of environmental risks knowledge and the adoption of sustainability values in business and household practices.

A complicating factor here is the capacity of households to acquire sustainable materials as inputs and then sustainably combine them. For instance, many low-income households cannot afford the higher cost of adopting sustainable lifestyles.

Moreover, unequal access to sustainability infrastructure and knowledge (e.g., education or information) also inhibits the emergence and performance of sustainable CE practices. These barriers fall under an environmental justice perspective and relate to enhancing the capacities of households and businesses to perform sustainable CE practices when carrying out daily activities.

4.3 Outputs and Outcomes in Sustainable CE Practice Performance

While sustainability outcomes are often produced by acquiring and combining sustainable inputs, they could also result from unsustainable processes. However, sustainability outcomes arising from unsustainable processes may appear to be fleeting or have less impact on broader sociomaterial CE practice constellations. This, therefore, suggests the importance of sustainably transforming socio-economic processes to create lasting outcomes. Furthermore, evaluating a sustainability outcome in CE research and policy should examine its multiple forms and scale of impact. For example, sustainable outcomes could include novel or upcycled materials and technologies, new or enhanced skills or procedures, and new or refined principles, values, and rules. Regarding the scale of impact, sustainability outcomes may vary by space (e.g., household space versus community or organisation versus industry) or period (e.g., short-term versus long-term).

The preceding description of the sustainability aspects of CE practices further emphasises the need for a proper conceptualisation of sustainable CE practices in research and policy. While performing some CE practices may result in sustainable outcomes, such as extending the material lifecycle, those CE practices may not intrinsically be sustainable. A suggested conceptualisation is that a sustainable CE practice follows processes and generates outcomes that align with sustainability principles, such as transforming materials into new products or utilising services that reduce material use. Therefore, as applied in Shittu (2022), researchers and policymakers must develop frameworks that zoom into CE practices to examine the intricate performance processes and zoom out to address the impact of interconnection among CE practices.

5 Innovation by Combination and Resourcefulness in Socio-technical Systems

Although many low-income households lack access to socio-economic resources, they combine low-cost materials with other practice elements in mundane, innovative ways to perform household practices. Examples include repurposing used or old materials from one household activity to another. This maximises value utilisation and extends the lifecycle of materials. Transferring personal items to significant others, often grounded in sentimental values attached to items, or sharing items with neighbours, similarly extends the lifecycle of materials but also reduces the financial burden on recipients and improves communality (Shittu 2023a). This is not to argue that these practices are sustainable per se but highlights the role socio-technical resourcefulness can and does play in rising consumption, income gap, disposability, wastage, and environmental degradation (Shittu 2020). It also highlights how existing practices inherently comprise elements associated with sustainable CE.

Therefore, transitioning cities to sustainable CE must also involve re-imagining new, radical, and disruptive practice configurations within existing markets, product and service design, technological innovations, infrastructure provision, community services, and policy landscapes. While it will remain essential to pursue domain-specific circular strategies and action plans (for instance, design for re-assembly in the construction sector), it is imperative for the barriers between and among domains (e.g., industries, business scales, knowledge domains) also to enable trans-local, multidisciplinary, and multi-stakeholder circular solutions. In this case, the transition challenge can be summarised thus: optimising socio-technical and socio-economic resourcefulness by innovatively combining existing approaches in new, sustainable, and circular ways.

Innovation by combination addresses two central issues of sustainable CE transitions: diverting existing materials away from the environment and creating new products and services that embed sustainability and circularity principles. Several examples of innovation by combination already exist, although currently at a marginal and premium rate. According to Heleven (2010), innovation by combination strategies includes (Fig. 4):

(1) designing products with multiple functionalities such as in fashion, construction, transportation, mobile technology, and household items;
(2) introducing new products to the market by using established brands such as in-home design and fast-moving consumer goods;
(3) conscious design of packaging with additional functions such as in-home appliances and food packaging;
(4) enabling user-led temporary combinations such as in spatial design;
(5) subsuming multiple products into each other as a nested design, such as in transport technology, mobile technology, and household appliances;
(6) innovative combinations of products and services, such as in mobile technology and product-as-service;
(7) integrating multiple products and services into one product or service (e.g., old and new materials, all-in-one or co-branding) such as in hybrid energy systems;
(8) oxymoronic design of products and services such as in virtual reality, home office and biotechnology, and
(9) adjectival innovations in industries (e.g., wildlife tourism, agritourism, and backpack tourism.

Notably, while innovations are essential in aiding socio-technical transitions, they should be developed in the context of sociomaterial practices. That is, effective and

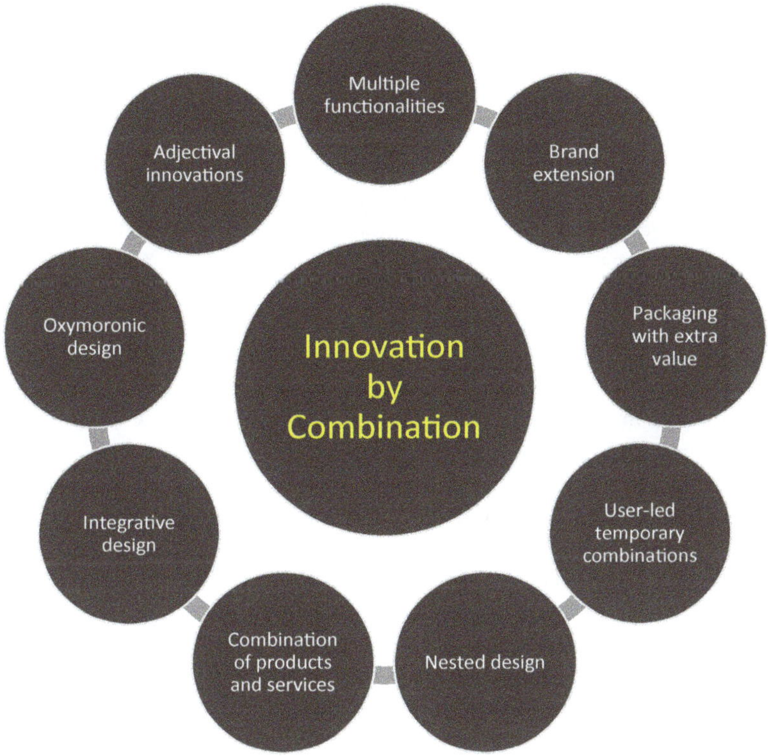

Fig. 4 Innovation by combination strategies. *Source* Authors' design following Heleven (2010)

sustainable innovations are those that consider the complexity and interconnectedness of existing systems when integrated. The current approach to introducing new CE materials or innovations (e.g., technologies, products, etc.) into society assumes that embedding materiality with sustainability elements is enough to reconfigure existing practice complexities. This reflects a transition dynamic grounded in (presumed) market efficiency and a primarily exogenously driven reconfiguration of daily activities in response to technological change. Although materials shape practices through their corporal, spatiotemporal, and functional dimensions (Shittu 2023a), a critical insight from practice theory is that daily activities are also shaped by other practice elements such as competence and meanings. A sustainable material not accompanied by sustainable skills and meanings may become ineffective within the context of a CE practice.

Moreover, no practice exists in isolation, and practitioners often prefer the practice that provides the highest value in a particular context. In such cases, a sustainable CE practice integrating innovations may become less effective if the attached values do not outweigh those of an unsustainable alternative or are impeded by other socio-economic factors. For instance, although some low-income households are keen on sorting their plastic waste, they are inhibited by the lack of control over their immediate environment (Shittu et al. 2021). This insight also speaks to the planning strategies across the global north that aim to increase residential density patterns and multi-unit dwellings in cities such as Melbourne (Chhetri et al. 2013; Motazedian 2017). Such planning strategies may reduce personal and shared spaces in residential areas and impact household sustainable CE practices, even for high-income households. Urban planning approaches for developing a compact city alongside transitioning to sustainable CE cannot just focus on infrastructure that aids the flow of people but also the flow of materials and their implications for urban household practices.

The preceding point extends to a more general point—creating an enabling environment for integrating sustainable innovation in CE practices. This involves facilitating sustainable infrastructural accessibility, values adoption, and skills adoption among households and businesses (Shittu 2023b). In essence, the successful transition of cities is predicated on cooperation among stakeholders and a combination of approaches across all spheres of the socio-technical system to produce novel processes and practices. Sustainable CE is a solution that requires different implementation strategies for each city by relying on local resources and networks. This necessitates promoting mutually beneficial and ecologically responsible partnerships in design, governance, markets, and communities (Shittu 2023b).

6 Conclusion and Policy Insights

Theoretical approaches that focus on individual behaviours or market forces often ignore the complex interaction of sociomaterial arrangements with daily activities and its impact on the transition of cities to sustainable CE. However, practice theory provides a framework to zoom into a specific activity in a domain and analyse its connection with complex systemic processes. For instance, findings from a doctoral study show that plastic's functions in household activities through its unique physical features and spatiotemporal interaction render it indispensable to households (Shittu 2022). Therefore, this creates complications for sustainable intervention programmes looking to eradicate or reduce plastic use in cities, among other plastic-related CE strategies.

Moreover, the importance of achieving sustainability and the CE in the city, national and global discourses has emphasised the role of understanding and embedding sustainable CE practices into societal processes. This implies developing and implementing strategies that aid in reorganising the internal processes of existing practices, substituting unsustainable practices, and reconfiguring practice complexities. However, such strategies must involve a more nuanced analysis of practices. This chapter argues that researchers, policymakers, businesses, and other sustainability stakeholders must be perceptive in evaluating the connectivity between CE practices, the influence of skills and meanings in CE practice performance, and creating lasting sustainability outcomes.

Sustainability meanings and values are anchored in materials. Therefore, a systemic realignment to sustainable CE requires new and innovative combinations of materials and technologies that aid CE practices and business models. Some 'innovations by combination' have been identified to include brand extension, user-led temporary combinations, and a combination of products and services. Examining sustainability considerations and promoting innovation by combination are two primary tools for addressing environmental justice issues and accelerating multi-stakeholder collaboration towards sustainable CE transitions.

Advancing the applied and institutional capabilities for developing sustainable CE solutions requires more detailed analysis and evidence on the impact of materiality on social formation in social sciences. Furthermore, sustainability strategies must be inclusive and address environmental justice issues. Finally, for design, it is essential to be conscious of how goods, services, and innovations would impact sustainable CE practices among consumers or households.

Drawing on the practice perspective, Table 1 recommends strategies for accelerating sustainable CE transitions in policy, research, and practice. Each recommendation is ranked based on its importance or significance for research, policy, and practice. Importance or significance level refers to the capacity of each domain to act directly on and achieve the corresponding strategy.

Table 1 Summary of study recommendations for accelerating sustainable CE transitions in research, policy, and practice

S/N	Recommendations	Theory/research	Policy	Design/practice
1	Sustainable CE transition strategies must be well-coordinated, trans-local, holistic, and integrative	Moderate relevance	High relevance	High relevance
2	Collaborate to sustainably redesign and reconfigure elements of existing practices, including materials (e.g., infrastructures), meanings (e.g., guidelines, rules, and values), and competencies (e.g., skills and processes)	High relevance	High relevance	High relevance
3	Scale up, sustain, and guide the transformation of sustainable CE solutions by leveraging communities of practice	Moderate relevance	High relevance	High relevance
4	Broadening of CE policy, practice, and research focus to impact practices as the underlying building blocks of social structures	High relevance	High relevance	High relevance
5	Adopt comparative analysis and systems thinking to address practice complexities during an intervention	High relevance	High relevance	High relevance
6	Further development of a theoretically informed understanding of materiality and material agency beyond anthropocentric perspectives	High relevance	Moderate relevance	Moderate relevance
7	Urban sustainability solutions should be designed to address environmental justice issues, e.g., by being inclusive, simple, affordable, and efficient	Moderate relevance	High relevance	High relevance
8	Adopt strategies that optimally integrate downstream sustainable CE solutions (e.g., recovery, recycling, and waste to energy) with upstream sustainable CE solutions (e.g., designing alternative sustainable materials)	High relevance	High relevance	High relevance
9	Investigate various sociomaterial combinations that can facilitate sustainable CE practices and design policies that optimise them in a circular economic system	High relevance	High relevance	Moderate relevance

(continued)

Table 1 (continued)

S/N	Recommendations	Theory/ research	Policy	Design/ practice
10	Engage in a more nuanced analysis of sustainability considerations in CE practices (e.g., in households and businesses)	High relevance	High relevance	High relevance
11	Optimise resourcefulness in socio-technical systems through innovation by combination	Moderate relevance	Highly relevant	Highly relevant

Source Shittu (2022)

Acknowledgements The authors express gratitude to the reviewers for their constructive and insightful feedback on the early versions of the chapter. The doctoral research discussed in this chapter was supported by Swinburne University of Technology, Australia, through the Swinburne University Postgraduate Research Award (SUPRA) and a Travel Grant from the Centre for Urban Transitions, Swinburne University.

References

Buhl J, Liedtke C, Teubler J, Bienge K, Schmidt N (2018) Measure or management? Resource use indicators for policymakers based on microdata by households. Sustainability 10(12)
Cai BM, Liu BB, Zhang B (2019) Evolution of Chinese urban household's water footprint. J Clean Prod 208:1–10
Chhetri P, Han JH, Chandra S, Corcoran J (2013) Mapping urban residential density patterns: compact city model in Melbourne, Australia. City Cult Soc 4(2):77–85
Davies AR, Doyle R (2015) Transforming household consumption: from backcasting to HomeLabs experiments. Ann Assoc Am Geogr 105(2):425–436
Heleven M (2010) 37 ways for innovation by combination. Slide Share. https://www.slideshare.net/marcnewshoestoday/37-ways-for-innovation-by-combination. Accessed 19 Apr 2022
Herrmann M, Brumby D, Oreszczyn T (2018) Watts your usage? A field study of householders' literacy for residential electricity data. Energ Effi 11(7):1703–1719
Ibrahim MA (2020) Modeling of risk for improper sorting of waste at Recycling Centers. Waste Manag 102:550–560
Jaeger-Erben M, Offenberger U (2014) A practice theory approach to sustainable consumption. GAIA-Ecol Perspect Sci Soc 23(s1):166–179
Kuijer L (2014) Implications of social practice theory for sustainable design. PhD thesis, Delft University of Technology. https://www.researchgate.net/profile/Lenneke_Kuijer/publication/266247132_Implications_of_Social_Practice_Theory_for_Sustainable_Design/links/55d355a008ae0a3417226651e/Implications-of-Social-Practice-Theory-for-Sustainable-Design.pdf. Accessed 1 Mar 2019
Lavelle MJ, Rau H, Fahy F (2015) Different shades of green? Unpacking habitual and occasional pro-environmental behavior. Glob Environ Chang 35:368–378
Lee JD, Bahl A, Black GS, Duber-Smith DC, Vowles NS (2016) Sustainable and non-sustainable consumer behavior in young adults. Young Consum 17(1):78–93
Marres N (2011) The costs of public involvement: everyday devices of carbon accounting and the materialization of participation. Econ Soc 40(4):510–533
Motazedian A (2017) The microclimatic interaction of a small urban park with its surrounding urban environment: case study of projected impacts of urban densification focusing on heat events in Melbourne. PhD thesis, Monash University, Melbourne

Nicolini D (2009) Zooming in and out: studying practices by switching theoretical lenses and trailing connections. Organ Stud 30(12):1391–1418
Nyström S, Dahlberg J, Edelbring S, Hult H, Dahlgren AM (2016) Debriefing practices in interprofessional simulation with students: a sociomaterial perspective. BMC Med Educ 16(1):1–8
Öztekin EE, Gaziulusoy İ (2020) Co-positioning design for sustainability transitions, practice theory and transitions theories: towards dialogue and collaboration. J Des Res 18(3/4):196–223
Paddock J (2017) Household consumption and environmental change: rethinking the policy problem through narratives of food practice. J Consum Cult 17(1):122–139
Pocol CB, Marinescu V, Amuza A, Cadar R, Rodideal AA (2020) Sustainable vs. unsustainable food consumption behaviour: a study among students from Romania, Bulgaria and Moldova. Sustainability 12(11)
Reckwitz A (2002) Toward a theory of social practices: a development in culturalist theorizing. Eur J Soc Theory 5(2):243–263
Schatzki T (1996) Social practices: a Wittgensteinian approach to human activity and the social. Cambridge University Press, Cambridge
Shittu O (2020) Emerging sustainability concerns and policy implications of urban household consumption: a systematic literature review. J Clean Prod 246(119034)
Shittu O (2023a) 'Almost everything in the house now is plastic': foregrounding plastic materiality in household routines and practices. Sociol Res Online. https://doi.org/10.1177/13607804211034887
Shittu O (2023b) Grassroots strategies for environmental governance and circular cities: lessons from Lagos and Melbourne. In: McCormick K, Evans J, Palgan YV, Frantzeskaki N (eds) A research agenda for sustainable cities and communities. Edward Elgar Publishing
Shittu O (2022) Transforming household practices for enabling urban transitions to circular plastic economies: a practice perspective. PhD thesis, Swinburne University of Technology, Melbourne
Shittu O, Nygaard C, Bailey A (2021) Sustainability practices and materiality: transforming plastic consumption in urban households. Manuscript submitted for publication
Shove E, Pantzar M, Watson M (2012) The dynamics of social practice: everyday life and how it changes. Sage, London
Shove E, Watson M, Hand M, Ingram J (2007) The design of everyday life. Berg, Oxford
Sole T, Wagner C (2018) Understanding domestic fuel use practices in an urban township. Build Res Inf 46(2):220–230
Strengers Y, Maller C (2014) Social practices, intervention and sustainability: beyond behaviour change. Routledge, London
Strengers Y, Nicholls L, Maller C (2016) Curious energy consumers: humans and nonhumans in assemblages of household practice. J Consum Cult 16:761–780
Warde A (2005) Consumption and theories of practice. J Consum Cult 5:131–153
Wenger-Trayner E, Wenger-Trayner B (2015) Communities of practice: a brief introduction. https://wenger-trayner.com/introduction-to-communities-of-practice/. Accessed 25 Aug 2021
Yoada RM, Chirawurah D, Adongo PB (2014) Domestic waste disposal practice and perceptions of private sector waste management in urban Accra. BMC Public Health 14(1):1–10
Zorell CV (2020) Nudges, norms, or just contagion? A theory on influences on the practice of (non-) sustainable behavior. Sustainability 12(24)

Addressing Psychological Needs in Designing for a Sustainable Circular Economy

9

Christian Wölfel and Michael Burmester

1 Challenges of Designing for Sustainable Development

1.1 Designing for Sustainability and Circular Economy

Designing for sustainability has already gone a long way (cf. Chap. 2 by Tischner in this book). Theoretical frameworks have been developed as well and many design proposals have been made. The underlying intention was and is to allow sustainable consumption and behavior. Early approaches concentrate on ecological sustainability; subsequently, this focus has been integrated with more attention to practical economic and social aspects.

More recently, approaches to designing for sustainability and circular economy comply with broader concepts of sustainability, as formulated in the UN's Sustainable Development Goals (Velenturf and Purnell 2021; Melles et al. 2022, Chaps. 1 and 2 of this book). The design discipline can contribute based on its competences, for example on dealing with complexity or allowing for participation in development processes. Systematic human-centered design can pay regards to the complex potentials and boundaries of changing of systems, organizations, regulations and business models in sustainable transition, but at the same time to existing technical constraints and requirements, market dynamics as well as consumer and user needs.

Recent sustainable design approaches have begun focusing on circular design as a contribution to the circular economy (e.g., Charter 2018, Desing 2021). A

C. Wölfel (✉)
Technische Universität Dresden, 01062 Dresden, Germany
e-mail: christian.woelfel@tu-dresden.de

M. Burmester
Hochschule der Medien Stuttgart, 70569 Stuttgart, Germany
e-mail: burmester@hdm-stuttgart.de

© The Author(s), under exclusive license to Springer Nature Singapore Pte Ltd. 2024
G. B. Melles and C. Wölfel (eds.), *Design for a Sustainable Circular Economy*, Design Science and Innovation, https://doi.org/10.1007/978-981-99-7532-7_9

sustainable circular economy pays regard to environmental quality, economic prosperity and social and individual wellbeing (Velenturf and Purnell 2021). In order to recognize these aims, efficiency, sufficiency and consistency strategies need to be implemented, resulting in "far-reaching changes in production, consumption and waste management, and collaboration and coordination to visualize implementation" (ibid.). Sustainable circular economy is hence characterized, not only, by long-term product use and circular product flows, e.g., through refurbishment (cf. Stahel 2019). While awareness campaigns and governance, norms, regulations and steering mechanisms can contribute to a wider implementation, the consensus is that human-centered design can play a key role in driving circular economy success (ibid., see, e.g., Chaps. 1 and 7 of this book).

There are various circular design solutions that are more sustainable than conventional ones, and that are at the same time attractive or even "desirable to a certain demographic" (Lofthouse and Prendeville 2018: 457). In general, it can be said that even on the level of material circularity, there is much room for improvement in many products and services that can be addressed by designers and product developers even with established approaches and methods. The design of more sustainable circular solutions can be accomplished by improving material circularity with less downcycling and less energy consumption. Although there is a critique that this allows only for incremental progress in sustainable transition, these limited and slow advances are still necessary.

Not all circular economy translates to sustainability (cf. Meindl 2021, Velenturf and Purnell 2021). Many markedly sustainable solutions turn out less sustainable upon systematic evaluation. Not all of these discrepancies are rooted in intended greenwashing. Furthermore, too often circularity is only accomplished on the lowest possible (read: worst) level of thermal/energetic recycling (read: burning the material, hence taking it from the cycle) (cf. Chap. 4 in this book). Hence, more recent models of sustainable circular economy strategies intend to promote sufficiency by more radical approaches such as Rethink and Refuse. Following the more recent R strategy models, not only more reuse, repurpose and repair of materials, components and products must be enabled, but also more rethinking and refuse is necessary (UNEP 2022).

Changing production and consumption patterns for more sustainability requires systems thinking. More sustainable solutions require not only rethinking and redesigning products as such, but product-service systems as a whole, and eventually accordingly changing industries (Tukker et al. 2017; Desing 2021; Velenturf and Purnell 2021). Product-service systems are understood as a complementary blend of tangible and nontangible assets addressing user requirements (Goedkoop et al. 1999). Both benefits for people and economic value can be generated through interwoven (tangible) products and (intangible) services (Boehm and Thomas 2013). Decoupling economic value creation and resource consumption, the assumption of less material use in physical products, thus increasing resource efficiency, the motivation for low maintenance and repair cost through longer lasting products, or the monitoring and lifecycle analysis using digital twins are considered promising to contribute to sustainable circular economy and sustainable development (Schmidt-Bleek 1998; Pieroni et al. 2016).

Yet remaining are challenges of possible negative side-effects and rebound effects such as increased consumption or increased wear of shared goods (e.g., Zink and Geyer 2017; Makov and Vivanco 2018). To avoid such negative effects, product-service systems must be developed specifically to meet the requirements of sustainable circular economy (Blüher et al. 2020; Pieroni et al. 2017). If implemented accordingly, product-service systems design can contribute to sustainable development and transition (Joore and Brezet 2014).

1.2 Barriers to Success of Sustainable Circular Economy

Despite a good understanding on the necessities and extensive efforts in promoting circular economy, independent reports suggest the success has stagnated for several years at a market share still around 10% (e.g., Morató et al. 2019; de Wit 2022). Discussions on the reasons for this often look at governance and producer perspectives. Other analyses identify the lack of consumer acceptance as one major challenge in circular economy (Camacho-Otero et al. 2018).

This can be caused by too much sudden change for customers, resulting in their refusal. Generally, if products and interactions are too novel, too different or too unusual, there is often a lack of acceptance among consumers and users (Hekkert et al. 2003; Oehme 2011, Blijlevens et al. 2012) which is often unexpected by designers who are generally much more open to change in the domains they are designing for (cf. Oehme 2011). Such lack of customer acceptance of novel solutions has also been identified in the context of circular design solutions. Hence, it is the challenge to deliver the Most Advanced Yet Acceptable (cf. Hekkert et al. 2010) applies also to sustainable circular economy. Yet, approaches such as design for disassembly and design for repair are well known and yield the potential to increase sustainability in rather conventional settings even without putting too much change on consumers and users. Other than the mere refusal to too much change among many customers, there are concerns of circular economy, e.g., in terms of reliability and quality of recycled materials or the trustworthiness of sharing economy models (e.g., Wallner et al. 2022; Kuah and Wang n/d). Evaluations of circular solutions often show limited acceptance and suggest or show raised acceptance due to more transparent communication and education of improved sustainability. To some degree, such limitations can be overcome over the course of time by regulations, and younger generations getting mature and getting used to it. Beyond such rather general proposed solutions, further strategies have been investigated that address specific acceptance barriers. For example, Wallner et al. (2022) provide a hands-on physical product design guideline that increased acceptance by reducing the contamination concerns of customers in the field of refurbished personal care products.

Lofthouse and Prendeville (2018), Velenturf and Purnell (2021) and others draw another perspective on how design can contribute to the acceptance of sustainable circular solutions. According to them, there is a relevant potential in participatory design. Involving and engaging users/customers in the development processes can lead to a better fit of customer demands and product-service offering.

However, methodological challenges are seen on how to include diverse stakeholder perspectives for co-creating sustainable output (Buhl et al. 2019; Kagan et al. 2020). Furthermore, there is still the problem of the intention-behavior gap in the context of circular solutions (cf. Camacho-Otero et al. 2018). Customers may contribute their opinions, beliefs and intentions in co-design and participatory design settings (Sanders and Stappers 2014). However, customers are found to make their actual consumption decisions independently from their original intentions (after all, this can be in favor of energy-efficient offerings; cf. Khor and Hazen 2017).

Also, even people who aim to act responsibly often make decisions to the disadvantage of sustainable development (Carrington et al. 2014). Decision-making is a complex but most often holistic process that is also characterized by unconscious simplification and prioritization and that is always influenced by mood, affects and emotions, no matter if people are aware of this and no matter how rational people consider themselves. Affects can influence the representation of the individual decision problem as well as the decision-making process, which is both relevant in the context of consumer behavior. For example, if there is (the feeling of) the need of a sudden solution, if customers feel they deserve a reward, e.g., for a week of hard work or if they just do not systematically weigh up arguments for any other of a multitude of possible reasons, decisions may drastically differ from the ones made based on original intentions and careful, informed and reflected considerations.

Based on a review of 111 studies, Camacho-Otero et al. (2018) identified seven major themes of factors that promote or prevent the acceptance of circular design solutions among customers that widely align with our above analysis:

- personal characteristics (of the customers, such as desire for change),
- product and service offering (such as product quality),
- knowledge and understanding (including information about services),
- experience and social aspects, risks and uncertainty (trust, disgust, newness, other risks),
- benefits (economic, environmental, social),
- other psychological factors (attitudes, norms, habits, values and perceived behavioral control).

As can be seen, only a small part of these factors is determined by the design process. A larger part of the factors can be potentially considered and recognized by design—if they are understood well enough and are considered despite their complexity, lack of transparency and their inconsistency. To a certain degree, some factors that promote or prevent the acceptance of sustainable circular solutions can be increased by education, regulation, nudging and other extrinsic means. In goodwill but ignoring for example relevant psychological factors, designers and sustainable entrepreneurs too often aim to convince their customers to follow what they consider to be the right solutions. Camacho-Otero et al. (2018) derive from their systematic review that:

> Most of the literature on the circular economy seems to focus on the production side, exploring circular business models, strategies to develop circular value propositions, and the benefits of such models. Less attention seems to have been paid to how consumption and consumers would affect or be affected by the circular economy. (no page).

Discussions on limited acceptance of novel sustainable circular solutions often focus on objective values of offerings and the attitudes of people in relation to their knowledge about circularity or sustainability (see above). To some degree, as suggested in such studies, improved product quality as well as appropriate communication and education will lead to changes in attitude and hence increased acceptance. Regulations will affect offerings on the markets, which will gradually diffuse to the knowledge, attitudes and habits of customers. These are two examples of (long-term) changes of human factors in the context of circular economy. Situations will also change with time, as may other factors. However, in short-term, human factors such as norms, beliefs, expectations, psychological needs and hence consumer behavior cannot easily be changed by design or regulations. There are approaches such as "design for behavioral change" that might seem to disprove the argument made here. However, in design for behavioral change there is no change (or even designing) of human factors. It is rather a productive acknowledgment and operationalization of existing human psychological factors in order to promote desirable conduct. The human side of human-product interaction, experiencing, resulting judgment and behavior will to a large part not be subject to change. But it has an important impact on conduct and consequences. Hence, it must be understood and considered when designing successful sustainable circular solutions. Not only is it necessary to put humans (users, customers) in the focus of sustainable circular design (cf. e.g., Selvefors et al. 2019) but more specifically their experience and behavior, and their psychological needs (Camacho-Otero et al. 2018). This can be accomplished by a new human-centered design approach that systematically integrates these psychological dimensions.

2 Designing for Positive Experiences and Wellbeing

2.1 Positive Experiences Through the Fulfillment of Psychological Needs

In the earlier days of industrial design and later on in the emerging field of human–computer interaction (HCI), a central problem in the design of physical and digital products was making them easy to use. Among the first approaches were to use physiological ergonomics as well as the flourishing theory building in cognitive psychology and to compile them to a design approach called user-centered design to ensure ease of use (Norman 1986). After having published the usability definition in the ISO Standard 9241-11 in 1998 (Bevan et al. 2015), the most important design approach human-centered design was derived and published for the digital domain as ISO standard 13,407 in 1999 (ISO 13407 1999). The focus was to design systems with good usability ensuring effective, efficient and satisfying use. The updates of the standard in 2010 and 2019 represent only few changes (ISO 9241-210 2019). It is a stable approach, which was very successful, because several innovation approaches use elements of human-centered design such as design thinking (Brown 2008) or lean start-up (Ries 2011). Recently, human-centered design and its psychological foundations are used also in non-digital disciplines such as organization development (ISO 27501 2019), sustainable tourism (Font et al. 2018), development of health services (Adam et al. 2020) and psychotherapy (Lyon et al. 2020).

The more recent construct of 'user experience' provides an increasingly important extension to the original approach. Similar developments have been seen in digital and physical product design (Uhlmann et al. 2016); eventually, both streams widely overlap in terms of design practice as well as theories and methodologies and finally join in the context of product-service systems design.

A first perspective on user experience has been defined that yields a broader view on usability acknowledging subjective aspects such as different kinds of reactions of the users toward a system, a broader temporal perspective (use from moment to moment, reflection of the use, expectations regarding the next use) and taking into account the whole user journey and its several touchpoints of the user with products and systems of a company (ISO 9241-210 2010; Law et al. 2009). The problem with this broad definition was that firstly there was no clear explanation of what is driving the essence of an experience and secondly what a positive user experience is.

- The significant step has been done with the definition of Hassenzahl stating that the core attribute of user experience is an evaluative feeling and positive user experience is the consequence of fulfilling psychological needs (Hassenzahl 2008; Hassenzahl et al. 2010).

After this important step, several studies (Hassenzahl et al. 2010; Tuch et al. 2016, 2013; Tuch and Hornbæk 2015) showed that (Burmester et al. 2017).

- emotions shape the character of experiences,
- in most cases, negative emotions are caused by usability problems, i.e., usability is a hygiene factor,
- positive experiences are indeed the consequence of the fulfillment of psychological needs,
- interestingly, positive emotions have twice the impact on user experience compared to negative ones in technology experiences.

Hassenzahl (2010) and Diefenbach and Hassenzahl (2017) used eight of ten psychological needs identified by Sheldon et al. (2001). Desmet and Fokkinga recently reviewed six models of psychological needs and compiled them to a new model with 13 fundamental psychological needs and 52 sub-needs (Desmet and Fokkinga 2020a). The fundamental needs are autonomy, beauty, comfort, community, competence, fitness, impact, morality, purpose, recognition, relatedness, security and stimulation. All fundamental needs are explained by a definition and four sub-needs. Autonomy for instance is defined as "Being the cause of your actions and feeling that you can do things your own way, rather than feeling as though external conditions and other people determine your actions." (Desmet and Fokkinga 2020a, p. 9 table 3). The four sub-needs associated with the fundamental need autonomy are freedom of decision, individuality, creative expression and self-reliance (2020a, p. 9 table 3). Different from other need models like Sheldon et al. (2001), they make a difference between the strong emotional bonds of the need relatedness described as "Having warm, mutual, trusting relationships with people who you care about, rather than feeling isolated or unable to make personal connections" (2020a, p. 9 table 3) and broader social relationships represented by the need for community which is defined as "Being part of and accepted by a social group or entity that is important to you, rather than feeling you do not belong anywhere and have no social structure to rely on" (2020a, p. 9 table 3). The full model description can be found in the above-mentioned publication. An illustrated description of the needs can be found in a small booklet which can be used for design work (Desmet and Fokkinga 2020b).

2.2 A Vision for a Successful Circular Economy Through Design for Positive Experience

Designing products for circular economy means to rethink typical quality criteria. Traditional technology focuses very much on novelty promising new functionality, new designs, etc. The corresponding intensive positive experience is the WOW experience. This experience is characterized by the emotions 'fascination' and 'joy' (Kulzer and Burmester 2018) and the most salient psychological need fulfilled is 'stimulation' (Kulzer and Burmester 2020), i.e., experiencing something new. But this way of attracting customers to buy a product means that new technical features and products must always be developed. But this means that it is associated with high resource consumption.

We would like to explain how the design of products must be changed when conducting circular economy. We will do this using the example of sharing economy of power tools, i.e., drilling machines, cordless screwdrivers and jigsaws. We will use three different insights from the research on positive user experience and design for wellbeing.

1. As described above, design for **fulfillment of psychological needs** is of central importance. There we will use several psychological needs, but two will be of

central importance, which are 'competence' and 'community'. Competence is a very important psychological need; it is laid out in the self-determination theory of Deci and Ryan (2000) and proven for experiences during technology use (Burmester et al. 2017; Hassenzahl et al. 2010; Tuch et al. 2016). Especially for tools, competence is an important psychological need, because tools are always used to achieve specific goals giving the users a feeling of effectivity and self-efficacy. In the model of psychological needs of Desmet and Fokkinga (2020a), the need of **'community'** is very interesting, because here feelings of being part of a group of like-minded people can be important.

2. The next design aspect is not a need, but a special type of experience incorporating different psychological needs—called **'product attachment'**. Product attachment means that the user develops a long-term emotional relationship to the product. This has been described by Mugge and colleagues (Mugge 2008; Mugge et al. 2010; Mugge and Schoormans 2006). At the first glance, this seems to be a strange concept for sharing economy, because products are just used for dedicated purposes and are not owned by the user. But an emotional bond to a product leads to more positive emotions and—this importance for sustainability—to protective behavior, like "to handle the product with care, to repair it when it breaks down, and to postpone its replacement" (Mugge et al. 2009, p. 468).
3. When designing for positive experiences, the **temporal dimension of experiences** is important (Hermosa Perrino and Burmester 2020; Pohlmeyer et al. 2009). The elementary experience is the moment-to-moment experience, e.g., during a usage activity. This experience is stored in the episodic memory, reflected and can be communicated to others (Hassenzahl et al. 2013; Rossman and Duerden 2019). Based on the personal or collective reflection of the experience, users develop expectations and anticipation toward the next experience with a product.

2.3 A Fictitious Use Case in the Power Tools Market

The global power tools market is worth more than 20 billion USD annually. With almost two-thirds of this market, the commercial/professional use is the larger share of this market (Placek 2019). Hence, it is important to design for the various stakeholders and their needs. In linear economy, power tool providers split their portfolios into residential and commercial products. Commercial tools are more reliable, and are designed to be used for a longer period. However, commercial products are also distributed through DIY stores and widely used by laypeople. From a consumer perspective on the products, this separation is not as clear as from the manufacturer's point of view.

Usually, more resources are consumed to produce commercial tools to be used much more heavily than residential ones. However, once bought and used for a short period, these professional power tools are unused for most of the time.

According to some statistics, conventional residential use power drills are said to be actually used less than 30 min a year or even in their entire lifetime (cf. Kessler 2015). This is contrary to the archetypical principles of circular economy: sufficiency, efficiency and consistency (Huber 1995). Too many resources are used, and too little use is made from it. But why do many laypeople choose professional power tools, even though they are more expensive and, based on the short period of actual use, do not yield more or better instrumental qualities? The reasons are to be assumed among the non-instrumental qualities that address psychological needs such as 'stimulation' or 'community' as explained above.

Circular economy approaches on power tools could be for example refurbishment or sharing economy business models. In both cases, the split into less and more durable and at the same time less and more resource-consuming portfolio may become obsolete. Development resources can be concentrated on one portfolio of durable, reliable power tools that can then be utilized to a maximum. Sharing the resources in order to gain more use from it seems adequate to contribute to sustainable development. Already more than 15 years ago, startup companies aimed at addressing this issue by, e.g., providing community-shared power drills. Most of these initiatives are long gone (Kessler 2015). To some degree, there were problems such as a lack of actual and immediate availability of the tools when needed. These are instrumental, hygienic factors that limit acceptance. However, the non-instrumental has also not been addressed adequately in order to deliver positive experiences and hence raise acceptance.

With more recent approaches on shared product-service systems, more mature digital platforms, learning algorithms and the learnings from makerspaces and tool libraries (cf. Ellen MacArthur Foundation n/d)—and with a design approach that aims at delivering positive experiences through the fulfillment of psychological needs—a new attempt on shared power tools could be made. A possible scenario of a commercially available sustainable product-service system could include mostly commercial use during the labor days and residential laypeople use on weekends and holidays. The non-instrumental quality of professional tools for laypeople could even be enhanced by the sharing of *professionally used* tools. As always, rebound effects such as reduced trust or even damage caused by (unintended) misuse of the gear by laypeople may yield rebound effects and reduce the acceptance for such business models among the professionals and need to be considered. In any case, there could be just one circular offering that is then tailored to the different target groups. In order to gain acceptance, the design of the product-service system must understand and address psychological needs and provide positive experiences for all relevant stakeholders.

In the following, we illustrate in a fictitious example on how sharing economy can be designed as a product-service system for power tools in order to enable positive experiences when using the service and the tools consumer/residential settings based on the above-made three criteria: 1. fulfillment of psychological needs, 2. novel ways of 'product attachment' and 3. paying regard to the temporal dimensions of user experience (Fig. 1).

The residential setting is illustrated by the case of Maria and Oliver:

Fig. 1 A fictitious example of a power tool sharing solution that aims at delivering positive experiences by fulfilling psychological needs of laypeople and professional workers

Maria and Oliver are in the process of renovating their house. They want to do this as much as possible on their own to save additional money for professional craftsmen and also enjoy the work. This renovation project is an isolated case for them. Therefore, they don't want to buy the power tools they need. In recent years they have had good experiences with renting the tools. They are both members of a sharing platform and both have an account.

One part of the renovation is, that they build a veranda with about 40 sqm. A company has built the basic construction and now Maria and Oliver want to lay down the wooden floor. That means that they need to have a drilling machine, two cordless screwdrivers and a circular saw. Maria logs on to her account and could see immediately a timeline of projects she had (positive memory of activity history). Some projects are shared with Oliver, so they have a joint history of the activities in these projects ('community': doing things together). Maria looks through the last activities. Oh yes, she and Oliver had a hard task together. Two weeks ago, they had to deconstruct a rotten wall of the old veranda. Both used a drill hammer borrowed from the platform. She can see that they used the drill hammers for four hours and the platform analyzed the drill hammer data and indicated that this was hard work and that only one other layperson had the same intensity that is rated comparable to professional work during the last three months ('competence').

The service offers a wide variety of readily available tools they can freely choose from ('autonomy'). As laypeople, they trust the power tool selection guide which suggests appropriate tools for the woodworking tasks ('stimulation', 'security'). The guide surprisingly suggests another tool to accomplish large parts of the work: a dedicated flooring nailer will allow them to go quicker, with less effort, and less energy consumption. The special nails are provided as part of the rental. A short tutorial video explains how easy, secure, and fun it is to use ('stimulation', 'security'). Maria ordered the chosen tools and the platform indicated that she and Oliver used the type of screwdrivers already one year ago ('product attachment': common history). The tools arrive by delivery service the next day. Provided is a certificate on the quality check, battery status and reliability guarantee ('security'). A visualization illustrates how much professional and non-professional work has already done and how much the tools are still capable of and how this relates to the resources used ('community', 'purpose']. They grab the tools and feel the weight and the material of the robust tools, they know that professionals rely on these ('competence', 'community').

Oliver and Maria started their work on the veranda floor and took out the devices. After switching on the screwdrivers its display shows the name of Maria and Oliver ('product attachment'). They start with the work. The tools give them feedback that they are used adequately: there is a confirmation if nails and screws are applied perpendicularly and tightly for long-lasting joints ('competence'). At the same time, this reduces wear of the moving parts. At hard to access corners, they decide to set the screws in a slanted angle. This is a bit harder to do and puts more stress on the tool. The tool interface informs about this but freely allows the operation ('autonomy'). The floor nailer and screwdrivers count all nails and screws placed by them. At the end of the day, they realize over 400 nails and screws! Again, they had the feeling that they did a tough, but good job! ('competence', 'community').

The commercial setting differs from the residential one, since contexts, tasks, people as well as the mechanisms to address psychological needs differ in everyday labor work. Further detailing would be necessary to elaborate on how the needs of the workers are met by the design of the product-service system in order to not only accept the system but also to foster proper use of the tools for maximum sufficiency, efficiency and consistency. Expert use modes could be activated that do not only correlate to the competencies and tasks of the professionals, but also provide specific positive experiences based on the specific needs in the corresponding contexts.

It has to be noted that in most professional cases, the labor workers are the actual users of the tools. Their intentions and motivations may differ, since they work for others (e.g., their company, their customers) and not for their own intrinsic benefit as laypeople do. Usually, business managers or buyers decide on the order of the tools, not only keeping in mind the business numbers but also anticipating the workers who should be enabled and motivated to do great jobs. The particular fictitious solution could be a leasing or other sharing model. It could be made very flexible for the company to get and pay for reliable tools of the right kind in the right number for their current jobs. This way, very small companies can also get a hold of specialized tools such as large core drills and also adapt their tool booking to order situations, holiday times and so on. This can not only increase flexibility and reduce cost, but it can also increase the sustainability of the circular solution regarding maximized use in relation to resource consumption. Depending on the size of the company, the specific tools and the service contract, specific tools can be assigned to specific workers. This can enhance the individual 'product attachment' and reduce potential rebound effects by increasing the individual care that is taken of the tools, which can extend the use period of the tools.

As the fictitious case illustrates, an acceptable sustainable circular solution can be designed as a product-service system that consistently focuses on delivering positive experiences by addressing psychological needs. In addition to a problem-solving perspective (e.g., set a screw to fix a wood blank), positive experiences are delivered that raise acceptance of novel solutions and motivate for sustainable use and good work. This can be achieved in the framework of established human-centered design processes. However, the particular focus on psychological needs and experiences must be employed consistently.

3 Implementing the Psychological Needs Approach to Sustainable Circular Design Processes

3.1 A Systematic Process to Address Complex Design Problems

Human-centered design has faced challenges and changes in the past. However, core principles remain valid, such as the human perspective, iterative processes and early prototyping. There is not only a long tradition in the design of addressing issues that we today see as aspects of design for sustainability, such as eco-design, social design and participatory design (cf. Chap. 2 by Tischner). There is also a long tradition of design methodology and methods research (Bayazit 2004) with recurring debates ranging from individual and independent artistry on one end to systematic problem-solving on the other (Jonas 2007). The diversity of the discipline, its schools, academic and non-academic institutions, and industrial practice results in a design landscape that is still characterized by a heterogeneity in terms of approaches, methods, methodology and cultures (Bobbe et al. 2016).

Long-established are the rather individualistic, arts and crafts-oriented, conceptual as well as more recent speculative design approaches. They prove their relevance in boldly forging ahead, in radically questioning current standards, in delivering speculative designs and in provoking and enabling debates and discourses. Even though they are being criticized for neglecting real-world constraints or for developing niche solutions that might have little or no immediate impact, they make a valuable contribution to the discourse on sustainable development within the discipline and beyond. In speculative design, hypothetic radical designs can be used as a 'critical medium' and 'catalyst for social dreaming' (Dunne and Raby 2013: iv), allowing for more concrete discussions on what futures (and correlating future solutions) are probable, plausible, possible and preferable. Niche solutions may have a limited initial impact but serve as important building blocks in multi-level sustainability transitions (cf. Geels 2011, Chap. 1 of this book).

Quite contrary in some regards is the continued and growing establishment of empirically rooted academic design research, education and practice in integrated product design engineering, industrial design engineering (De Vere et al. 2010), human–computer interaction (Holtzblatt and Beyer 2016; Rosson and Carroll 2003) and human-centered design approaches (ISO 9241-210 2019; Meyer and Norman 2020) as well as in design for positive experiences and wellbeing (Desmet and Pohlmeyer 2013; Diefenbach and Hassenzahl 2017). These disciplines tackle the seemingly incalculability of human behavior and experience, fuzzy front ends of new product design processes, the complexities of cyber-physical systems and product-service systems as well as technological and societal changes by academic standards. Their working styles, languages and cultures are widely compatible with engineering and business developers.

Their systematic approach, empirical foundation and yet holistic solution-oriented problem-solving make them the ideal partners to deal with complex systems design problems where human experiencing and behavior, (new) technology and necessary changes for sustainable development need to be addressed. These design disciplines are potentially capable of dealing with the exacerbated challenges of designing for a sustainable circular economy. The design tasks are often or always characterized by, not only,

- environmental, social and economic constraints and impacts,
- through the whole lifecycle,
- the complexity of product-service systems,
- being part of industrial, economic and socio-technical systems,
- underlying (changes on) a multitude of norms and regulations,
- expectations, demands and needs of a multitude of stakeholders,
- with initial lack of or incomplete data
- and a lack of inappropriate benchmarks.

Such multidimensional challenges can only be accepted by interdisciplinary teams that systematically collaborate with experts and knowledge from different domains. The development of sustainable product-service systems and the provision of such offerings involve more disciplines and partners than with a traditional product. In such a multi-stakeholder network, interaction and cooperation occur between companies, users and even public institutions (Luiten et al. 2001), and these interdependencies must be considered throughout the entire lifecycle of the Product-Service System.

Systematic design approaches can potentially cope with the complexity and heterogeneity of demands as described above. However, in the context of designing for largely acceptable, yet acceptable larger changes, it is further required to understand and acknowledge.

- the psychological needs,
- their fulfillment,
- the corresponding human experiencing,
- its influences on decision-making and behavior,
- of diverse and specific humans
- in diverse and specific contexts.

How can this be achieved and how can designers be enabled to develop corresponding circular solutions?

3.2 A Theory-Driven and A More Empirical Approach

The human-centered design process remains the core approach (e.g., ISO 13407 1999). As a supplement, the focus on psychological needs is implemented consistently, resulting in a design solution that not only allows for instrumental qualities (the offering helps to fulfill a task, to solve a problem) but also for non-instrumental qualities such as positive experiences through the fulfillment of psychological needs (Fig. 2). For the actual designing to fulfill psychological needs, methods are available that can be used by trained experts (Diefenbach and Hassenzahl 2017; Peters et al. 2020; Zeiner et al. 2017). A sound understanding of psychological needs and the possibilities of their fulfillment are necessary both in general and in the context of the markets and use scenarios. This requires not only expertise on psychological needs and designing for positive experience. In order to facilitate the potential of this approach, a pertinent mindset and culture is necessary in the organizations and development teams.

There are two basic approaches to design for positive experiences and for wellbeing. One approach is theory-driven. Designing for positive experiences and wellbeing by the use of psychological needs is the most often applied theoretical approach. This can be done by using need cards in design processes describing the attributes of the psychological needs (Desmet and Fokkinga 2020a; Diefenbach and Hassenzahl 2017; Hassenzahl n.d.). Desmet (2012) identified 25 positive emotions which can be used for product designs and integrated in design processes as Emotional Granularity Cards (Yoon 2015). Other authors have used theories of

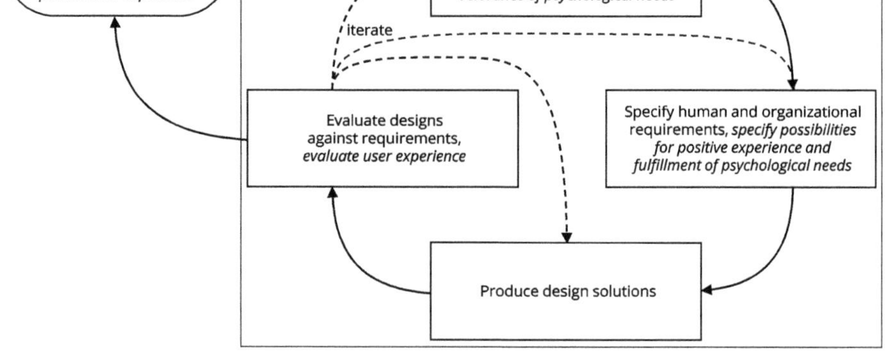

Fig. 2 The basic human-centered design process (ISO 9241-210 2019, upright text) adapted and extended to address psychological needs and positive user experience (text in italics)

positive psychology to derive design models for positive experiences like Positive Design (Desmet and Pohlmeyer 2013) or Positive Computing (Calvo and Peters 2014; Peters et al. 2018). A second approach is more empirically oriented. Experience categories is a typology of collected positive experiences from a specific context of use, e.g., like work context (Zeiner et al. 2018) or cooking in kitchens (Zeiner et al. 2018). As a second mostly empirical approach, Klapperich and colleagues (2018) focus on collecting positive practices as a basis for design.

3.3 A Dedicated Mindset and Systems Thinking

Designing for positive experiences and wellbeing requires additional knowledge and a new mindset for design for products. Desmet and Hassenzahl (2012) described a tendency to design products in order to solve a problem. They call it 'problem-driven design'. For example, a route guidance system solves the problem of finding the way from a starting point to a destination during car driving, cycling or walking. Desmet and Hassenzahl further state that if the goal is to design for happiness, then possibilities for positive experiences have to be found. They refer to this as 'possibility-driven design'. For example, geo cashing is using the same technology (GPS) like route navigation systems, but the purpose is to create situations of fun, interest, company, etc., which is finding the cash in interesting and fascinating places, having fun together with friends or family and sharing the experiences on a social network.

When designing for positive experience, this shift in the mindset is a prerequisite. Peters and colleagues interviewed 20 designers using positive computing methodology. They wrote that it is necessary that the designers need a deep theoretical understanding of positive psychology and the special design methodology. Burmester and Laib (2019) collected experiences from design workshops and found that even with theoretical and methodology knowledge, very often even experienced designers fall back in the problem-solving habits in design.

Consequently, designing for a sustainable circular economy requires further systems and lifecycle thinking. Marwede and Scholz (2022) suggest "combine methods from life-cycle thinking, planetary stakeholder analysis as well as user-centric business design" in order to create sustainable circular product-service systems. There is the necessity and correlating methods for designing product-service systems as well as for lifecycle design and assessment (e.g., Cattaneo et al. 2019; Chou 2021; Sasanelli et al. 2019; Walzberg et al. 2021). The lifecycle approach helps to recognize the total impact of the whole system in its whole lifecycle and to inform decision-making accordingly throughout the design process. While the applicability of quantitative lifecycle analyses is limited due to the lack of reliable data in early stages of product-service systems design, qualitative approaches of lifecycle and systems thinking must complement these (Li et al. 2021; Song et al. 2021).

3.4 Adapted Methods to Address Psychological Needs in the Human-Centered Design Process

In order to further integrate needs-based experience design, established human-centered design methodology can be adapted and extended to pay regard to this approach (Fig. 2). In the early stages, persona method, user/customer journey maps and empathy maps are typically tailored to the markets, the use cases as well as to the requirements of the developers and their organizations. Specific regard to psychological needs can easily be implemented here (Vogel et al. 2019; Anke et al. 2022). Furthermore, specific tools are provided that address psychological needs and can be implemented through the whole design process starting in the analysis phase (Hassenzahl and colleagues 2010; Desmet and Fokkinga 2020a; Zeiner et al. 2018).

In the middle stages, the fulfillment of psychological needs for positive experiences must be prototyped accordingly. Priorities on the fidelity of prototypes and test scenarios must be adjusted in order to allow for the evaluation of the user experience. While instrumental qualities need to be delivered at a minimum level in order not to disappoint test users (hygiene factors), the fulfillment of psychological needs for positive experiences on non-instrumental qualities must be prototyped more faithfully. Figures 3 and 4 show professional gardeners' testing experience prototypes of gardening robots. In the background, staff of the development team are imitating the core technical functions of the robot, while the front-end delivers the experiences that are being evaluated in this test.

In the late stages of the development process, more refined prototypes are available and higher fidelity setups are available for expert and user tests. For this purpose, well-evaluated evaluation methods are available that allow for the assessment of the fulfillment of psychological needs and positive experiences and that are state of the art in systematic experience design. One example is the psychological needs scales questionnaire developed by Sheldon and colleagues (2001). This standardized method has been transferred and evaluated for experience design by Hassenzahl and colleagues (2010) and is being used in academic research and industrial practice of experience design (e.g., Vogel et al. 2019). For a detailed formative evaluation of the user experience based on the definition of user experience by Hassenzahl (2008), the valence method can be used (Burmester, 2013; Burmester et al., 2010).

4 Discussion and Outlook

In the chapter, we have shown that there is a limited acceptance of sustainable circular economy solutions. We follow the analyses that identify the lack of attention to psychological factors in consumer behavior as a relevant barrier in sustainable transition. Considering the complexity of design tasks in sustainable circular economy, we argue that only a systematic design approach that pays regard to

Fig. 3 A professional gardener is participating in a user experience evaluation of a novel sustainable gardening robot. Members of the development team imitate the technical functions. © Tobias Ritz

the instrumental as well as to the non-instrumental qualities is necessary. In particular, we recommend the needs-based experience design approach for designing sustainable circular economy solutions. The current state of the art on user experience and design for wellbeing provides appropriate theories, design frameworks and effective methods for analysis, design and evaluation. In addition, there is empirical evidence that the application of all this in design processes results in products and services offering opportunities for positive experiences (Hassenzahl et al. 2021; Laib et al. 2022, 2018). But research shows also that there are still challenges. Small and medium enterprises have still a strong need for more information and a better knowledge on user experience (Haspel and Burmester 2021). Applying theories and using methods for a positive user experience is still a big challenge even for experienced interaction and product designers (Burmester and Laib 2019). A solid theoretical and practical training is necessary for beginners and design experts when they start with design for positive experiences (Peters et al. 2020). The Special Interest Group 'The Positive X' of the German Usability and User Experience Professionals Association (German UPA) brings user experience professionals together to further develop design methods and tools as well as to overcome hurdles in the design for positive experiences. They analyzed that, on the

Fig. 4 Positive experiences are necessary to increase the acceptance of novel solutions: two professional gardeners enjoy the interaction with a novel product-service system during a user experience evaluation. © Tobias Ritz

one hand, companies show a high interest in design for positive user experience. On the other hand, they most often opt for traditional designs in day-to-day decisions about time and money for personnel training and especially when deciding which functions to implement in products. But it also seems difficult for customers to understand the benefits and the value of products that enable positive experiences (Hermosa-Perrino et al. 2021). Here, we see more need for research to overcome the identified hurdles for successful products and services offering possibilities for positive experiences at work and daily life.

But there are more challenges. One is to better understand the impact of positive experiences. For example, a study showed that extended feedback ('competence') for very simple tasks firstly creates positive experiences and secondly leads to more concentration and commitment in the task (Kohler et al. 2007). Such changing effects of non-instrumental design qualities over the use span are of particular interest in design for change, especially in sustainable transition.

There is also a theoretical perspective. The broaden-and-build theory describes that positive emotions lead to more cognitive flexibility and a better coping with negative experiences (Fredrickson 2001; Tugade and Fredrickson 2004). Thus, positive emotions support creativity and resilience. By far, most positive user experience studies are short-term studies, very often laboratory studies. In order to show long-term effects, experiential features are required in products or prototypes that can be integrated into everyday life or work. For the circular economy, this means that design for positive experiences must be integrated into innovation processes and then tested. Products and services resulting from these processes must be examined in long-term studies (e.g., for the sharing economy scenario described above) in order to validate the processes and gain further insight into the design for positive experiences in the context of circular economy.

The example given in this chapter is fictitious and has been given to illustrate the potential of the needs-based experience design approach for the success of a sharing economy offering in sustainable circular economy. A number of facets of this example are already implemented in different readily available services and can hence be evaluated individually. The system as a whole has still to be prototyped in a way that allows for a comprehensive evaluation of the user experience, of the fulfillment of psychological needs and of the overall acceptance of such solutions. While the power tools market might seem narrow in terms of its impact to sustainable transition, it can deliver valuable insights in broader acceptance of circular economy in a field, where today the lack of systematic regard to psychological aspects are barriers to more sufficiency, efficiency and consistency. It can be argued that similar effects may be expected in other areas that have larger environmental and social impacts such as transportation or home appliances. The power tools case allows for a rigorous empirical evaluation in order to identify fundamental potentials of the approach. These can then be transferred and further investigated in more complex systems such as sustainable circular transportation solutions.

References

Adam MB, Minyenya-Njuguna J, Karuri Kamiru W, Mbugua S, Makobu NW, Donelson AJ (2020) Implementation research and human-centred design: how theory driven human-centred design can sustain trust in complex health systems, support measurement and drive sustained community health volunteer engagement. Health Policy Plan 35(2), ii150–ii162. https://doi.org/10.1093/heapol/czaa129

Anke J, Wölfel C, Schmitt F, Petzold C (2022) INSPIRe – Entwicklung eines Gamification-Ansatzes zur Attraktivitätssteigerung der Infrastruktur durch spielerische Maßnahmen im Straßenverkehr. Project report, https://doi.org/10.25368/2022.67

Bayazit N (2004) Investigating design: a review of forty years of design research. Des Issues 20(1):16–29

Bevan N, Carter J, Harker S (2015) ISO 9241-11 revised: What have we learnt about usability since 1998? Lecture Notes in Computer Science (Including Subseries Lecture Notes in Artificial Intelligence and Lecture Notes in Bioinformatics), 9169:143–151. https://doi.org/10.1007/978-3-319-20901-2_13

Blijlevens J, Carbon C-C, Mugge R, Schoormans JPL (2012) Aesthetic appraisal of product designs: Independent effects of typicality and arousal. Br J Psychol 103:44–57. https://doi.org/10.1111/j.2044-8295.2011.02038.x

Blüher T, Riedelsheimer T, Gogineni S, Klemichen A, Stark R (2020) Systematic literature review—effects of pss on sustainability based on use case assessments. Sustainability 12 (17):S.6989. https://doi.org/10.3390/su12176989

Boehm M, Thomas O (2013) Looking beyond the rim of one's teacup: a multidisciplinary literature review of product-service systems in information systems, business management, and engineering & design. J Clean Prod 51:245–260

Bobbe T, Krzywinski J, Wölfel C (2016) A comparison of design process models from academic theory and professional practice. In: DS 84: Proceedings of the DESIGN 2016 14th international design conference. pp 1205–1214
Brown T (2008) Design thinking design thinking. harvardbuisness review, 10. Academico/Material Didatico/Bibliografia 2016
Buhl A, Schmidt-Keilich M, Muster V, Blazejewski S, Schrader U, Harrach C, Süßbauer E (2019) Design thinking for sustainability: why and how design thinking can foster sustainability-oriented innovation development. J Clean Prod 231:1248–1257
Burmester M (2013) Valenzmethode – formative evaluation der user experience. In Scherfer K, Volpers H (eds) Methoden der Webwissenschaft – Ein Handbuch. Bd. I Anwendungsbezogene Methoden, Vol 11, pp 141–160. LIT Verlag
Burmester M, Laib M (2019) Warumfällt das Positive so schwer? Beschreibung von Herausforderungen bei der Gestaltung einer positiven User Experience. Mensch und Computer 2019 – Workshopband, pp 269–273. https://doi.org/10.18420/muc2019-ws-287-01
Burmester M, Laib M, Zeiner KM (2017) Positive Erlebnisse und Wohlbefinden in Arbeitskontexten durch Gestaltung der Mensch-Computer-Interaktion. In: Brohm-Badry M, Peifer C, Greve JM (eds) Positiv-Psychologische Forschung im deutschsprachigen Raum – State of the Art. Pabst
Burmester M, Mast M, Jäger K, Homans H (2010) Valence method for formative evaluation of user experience. In: Proceedings of the 8th ACM conference on designing interactive systems - DIS, vol 10, pp 364–367. https://doi.org/10.1145/1858171.1858239
Camacho-Otero J, Boks C, Pettersen IN (2018) Consumption in the circular economy: a literature review. Sustainability 10:2758. https://doi.org/10.3390/su10082758
Calvo RA, Peters D (2014) Positive Computing – Technology for Wellbeing and Human Potential. MIT Press
Carrington MJ, Neville BA, Whitwell GJ (2014) Lost in translation: exploring the ethical consumer intention–behavior gap. J Bus Res 67:2759–2767
Cattaneo L, Cassina J, Petrucciani M, Terzi S, Wellsandt S (2019) Models, methods and tools for product service design. Springer Nature. https://doi.org/10.1007/978-3-319-95849-1
Charter M (ed.) (2018) Designing for the circular economy, 1st edn. Routledge
Chou J-R (2021) A TRIZ-based product-service design approach for developing innovative products. Comput Ind Eng 161:107608. https://doi.org/10.1016/j.cie.2021.107608
De Vere I, Melles G, Kapoor A (2010) Product design engineering–a global education trend in multidisciplinary training for creative product design. Eur J Eng Educ 35(1):33–43
De Wit M (2022) The circularity gap report 2022. Circ Econ. https://circularity-gap.world/2022#Download-the-report
Deci EL, Ryan RM (2000) The 'what' and 'why' of goal pursuits: human needs and the self-determination of behavior. Psychol Inq: Int J Adv Psychol Theory 11(4):227–268
Desing H (2021) Product and service design for a sustainable circular economy. Doctoral dissertation, ETH Zurich.https://doi.org/10.3929/ethz-b-000472519
Desmet PMA (2012) Faces of product pleasure: 25 positive emotions in human-product interactions. Int J Des 6(2):1–29
Desmet PMA, Fokkinga SF (2020a) Beyond Maslow's Pyramid: Introducing a typology of thirteen fundamental needs for human-centered design. Multimodal Technologies and Interaction 4(3):1–22. https://doi.org/10.3390/mti4030038
Desmet PMA, Fokkinga SF (2020b) Thirteen fundamental psychological needs. Delft University of Technology
Desmet PMA, Hassenzahl M (2012) Towards happiness: possibility-driven design. In: Zacarias JV, Oliveira M (ed) Human-computer interaction: the agency perspective. Springer, pp 3–27. https://doi.org/10.1007/978-3-642-25691-2
Desmet PMA, Pohlmeyer AE (2013) Positive design: an introduction to design for subjective well-being. Int J Des 7(3):5–19. http://www.ijdesign.org/ojs/index.php/IJDesign/article/view/1666/595

Diefenbach S, Hassenzahl M (2017) Psychologie in der nutzerzentrierten Produktgestaltung. Springer
Dunne A, Raby F (2013) Speculative everything: design, fiction, and social dreaming. MIT Press, Cambridge, Mass
Ellen MacArthur Foundation: how tool sharing could become a public utility: Toronto Tool Library and Makerspace. Circular Example. https://ellenmacarthurfoundation.org/circular-examples/how-tool-sharing-could-become-a-public-utility
Font X, English R, Gkritzali A (2018) Mainstreaming sustainable tourism with user-centred design. J Sustain Tour 26(10):1651–1667. https://doi.org/10.1080/09669582.2018.1491981
Fredrickson BL (2001) The role of positive emotions in positive psychology: the broaden-and-build theory of positive emotions. Am Psychol 56(3):218–226
Geels FW (2011) The multi-level perspective on sustainability transitions: responses to seven criticisms. Environ Innov Soc Trans 1(1):24–40. https://doi.org/10.1016/j.eist.2011.02.002
Goedkoop MJ, van Halen CJG, te Riele HRM, Rommens PJM (1999) Product service systems, ecological and economic basics. Report for Dutch Ministries of Environment (VROM) and Economic Affairs (EZ)
Hassenzahl M (n. d.) Needs Cards. Experience Design Tools. http://www.experienceandinteraction.com/tools/
Hassenzahl M (2004) The interplay of beauty, goodness, and usability in interactive products. Hum-Comput Interact 19(4):319–349. https://doi.org/10.1207/s15327051hci1904_2
Hassenzahl M (2010) Experience design: technology for all the right reasons. Morgan & Claypool. https://doi.org/.2200/S00261ED1V01Y201003HCI008
Hassenzahl M (2008) User experience (UX): towards an experiential perspective on product quality. In: Proceedings of the 20th International Conference of the Association Francophone d'Interaction Homme-Machine, pp 11–15. http://portal.acm.org/citation.cfm?id=1512717
Haspel C, Burmester M (2021) Knowledge and implementation of usability and user experience in small and medium-sized enterprises. In Pfleging B, Kern D (eds) Mensch und Computer 2021. ACM, pp 131–135. https://doi.org/10.1145/3473856.3474299
Hassenzahl M, Burmester M, Koller F (2003) AttrakDlff : Ein Fragebogen zur Messung wahrgenommener hedonischer und pragmatischer Qualität 1 Einleitung Grundannahmen und Vorarbeiten. In: Szwillus J, Ziegler G (ed) Mensch & Computer 2003: Interaktion in Bewegung. B. G Teubner, pp 187–196
Hassenzahl M, Diefenbach S, Göritz A (2010) Needs, affect, and interactive products – Facets of user experience. Interact Comput 22(5):353–362. https://doi.org/10.1016/j.intcom.2010.04.002
Hassenzahl M, Burmester M, Koller F (2021) User experience is all there is - twenty years of designing positive experiences and meaningful technology. J Interactive Media I-Com 20(3):197–213. https://doi.org/10.1515/icom-2021-0034
Hassenzahl M, Eckoldt K, Diefenbach S, Laschke M, Lenz E, Kim J (2013) Designing moments of meaning and pleasure. Experience design and happiness understanding experiences. Int J Des 7(3):21–31
Hekkert P, Snelders D, Van Wieringen PCW (2003) 'Most advanced, yet acceptable': typicality and novelty as joint predictors of aesthetic preference in industrial design. Br J Psychol 94:111–124. https://doi.org/10.1348/000712603762842147
Hermosa-Perrino, C., Burmester, M., Spohrer, A., Fink, V., & Zeiner, K. M. (2021). The Positive X-Warum klappt das eigentlich nicht? Über die Herausforderungen und Schwierigkeiten bei der Integration von Positive UX im. In E. Ludewig & T. Jackstädt (Eds.), Mensch und Computer 2021 – Usability Professionals. Gesellschaft für Informatik e.V. und die German UPA e.V. https://doi.org/10.18420/muc2021-up-157
Hermosa Perrino C, Burmester M (2020) Designing for temporal harmony: exploring the well-being concept for designing the temporal dimension of user experience. Multimodal Technol Interact 4(3):66. https://doi.org/10.3390/mti4030066
Holtzblatt K, Beyer H (2016) Contextual design: design for life. Morgan Kaufmann

Hornbæk K, Hertzum M (2017) Technology acceptance and user experience: a review of the experiential component in HCI. ACM Trans Comput-Hum Interact 24(5):1–30. https://doi.org/10.1145/3127358

Huber J (1995) Nachhaltige Entwicklung. Strategien für eine Ökologische und Soziale Erdpolitik; Sigma, Berlin

ISO 13407. (1999). Human-Centred Design Processes for Interactive Systems.

ISO 27501. (2019). The human-centred organization – Guidance for managers. Beuth Verlag

ISO 9241-11. (2018). Ergonomics of human-system interaction – Part 11: Usability: Definitions and concepts.

ISO 9241-210. (2010). Ergonomics of human-system interaction – Part 210: Human-centred design for interactive systems (Vol. 2010).

ISO 9241-210 (2019) Ergonomics of human-system interaction – Part 210: Human-centred design for interactive systems.

Jonas W (2007). Design research and its meaning to the methodological development of the discipline. In: Design research now. Birkhäuser Basel, pp 187–206

Joore P, Brezet H (2014) A multilevel design model: the mutual relationship between product-service system development and societal change processes. J Clean Prod. https://doi.org/10.1016/j.jclepro.2014.06.043

Kagan S, Hauerwaas A, Helldorff S, Weisenfeld U (2020) Jamming sustainable futures: assessing the potential of design thinking with the case study of a sustainability jam. J Clean Prod 251:119595

Kessler S (2015) The "sharing economy" is dead, and we killed it. Fast Company, p 14

Klapperich H, Laschke M, Hassenzahl M (2018) The positive practice canvas – gathering inspiration for wellbeing-driven design. In: Proceeding NordiCHI'18, September 29-October 3, 2018, Oslo, Norway. https://doi.org/10.1145/3240167.3240209

Kohler K, Niebuhr S, Hassenzahl M (2007) Stay on the ball! In: An interaction pattern approach to the engineering of motivation. Lecture notes in computer science (Including Subseries Lecture Notes in Artificial Intelligence and Lecture Notes in Bioinformatics), vol 4662. LNCS(PART 1):519–522. https://doi.org/10.1007/978-3-540-74796-3_51

Khor KS, Hazen BT (2017) Remanufactured products purchase intentions and behaviour: evidence from Malaysia. Int J Prod Res 55(8):2149–2162. https://doi.org/10.1080/00207543.2016.1194534

Kulzer M, Burmester M (2018) Outstanding UX - Eine systematische Untersuchung von Wow-Effekten. Tagungsband Der Mensch Und Computer

Kulzer M, Burmester M (2020) Towards explainable and sustainable wow experiences with technology. Multimodal Technol Interact 4(3):1–18. https://doi.org/10.3390/mti4030049

Laib M, Burmester M, Zeiner KM, Schippert K, Holl M-L, Hennig D (2018) Better together - Unterstützung des positiven Erlebnisses der Zusammenarbeit durch Softwaregestaltung. In: Brohm-Badry M, Peiffer C, Greve J, Berend B (eds) Wie Menschen wachsen - Positiv-Psychologische Entwicklung von Individuum, Organisation und Gesellschaft, pp 73–90 Pabst Science Publishers

Laib M, Haspel C, Stockinger C, Polanski-Schräder L, Dücker P, Voll K, Schiffrer P, Burmester M (2022) Wellbeing at work—emotional impact on workers using a worker guidance system designed for positive user experience. Multimodal Technol Interact 876(10):1–31. https://doi.org/10.3390/mti6100087

Law ELC, Roto V, Hassenzahl M, Vermeeren APOS, Kort J (2009) Understanding, scoping and defining user experience: a survey approach. In: Proceedings of the 27th international conference on human factors in computing systems, pp 719–728. http://portal.acm.org/citation.cfm?id=1518701.1518813

Li X, Wang Z, Chen C-H, Zheng P (2021) A data-driven reversible framework for achieving sustainable smart product-service systems. J Clean Prod 279:123618. https://doi.org/10.1016/j.jclepro.2020.123618

Lofthouse V, Prendeville S (2018) Human-centred design of products and services for the circular economy–a review. Des J 21(4):451–476

Luiten H, Knot M, van der Horst T (2001) Sustainable product-service-systems: the kathalys method. In: Proceedings second international symposium on environmentally conscious design and inverse manufacturing. second international symposium on environmentally conscious design and inverse manufacturing. Tokyo, Japan, 11–15 Dec. 2001: IEEE Comput. Soc, pp S.190–197

Lyon AR, Brewer SK, Areán PA (2020) Leveraging human-centered design to implement modern psychological science: return on an early investment. Am Psychol 75(8):1067–1079. https://doi.org/10.1037/amp0000652

Makov T, Font Vivanco D (2018) Does the circular economy grow the pie? The case of rebound effects from smartphone reuse. Frontiers in Energy Research 6:39

Morató J, Jiménez LM, et al (2019) Situación Y Evolución De La Economía Circular En España. Decembre 2021. https://cotec.es/observacion/economia-circular/f62c16db-5823-deb4-7986-a786e5c3401c

Marwede M, Scholz R (2022) Can actant-centric system analyses and sustainable value proposition methods be an approach to sufficiency? In Augsten A, Wölfel C, Frye A, Köck M (eds) Design × Sustainability: Materiality/Systems/Shared Prosperity, pp 98–101. https://doi.org/10.25368/2022.294

Mahlke S, Thüring M (2007) Studying antecedents of emotional experiences in interactive contexts. In: Proceedings of the SIGCHI conference on human factors in computing systems - CHI '07, 915. https://doi.org/10.1145/1240624.1240762

Meidl R (2021). Disentangling circular economy, sustainability, and waste management principles. Issue brief no. 07.29.21. Rice University's Baker Institute for Public Policy

Meyer MW, Norman D (2020) Changing design education for the 21st century. She Ji 6(1):13–49. https://doi.org/10.1016/j.sheji.2019.12.002

Mugge R (2008) Emotional bonding with products: investigating product attachment from a design perspective. in investigating product attachment from a design perspective. VDM Verlag Dr. Müller

Mugge R, Schifferstein HNJ, Schoormans JPL (2010) Product attachment and satisfaction: Understanding consumers' post-purchase behavior. J Consum Mark 27(3):271–282. https://doi.org/10.1108/07363761011038347

Mugge R, Schoormans JPL (2006) A longitudinal study of product attachment and its determinants. Adv Consum Res 7:641–647

Mugge R, Schoormans JPL, Schiffersteinb HNJ (2009) Emotional bonding with personalised products. J Eng Des 20(5):467–476. https://doi.org/10.1080/09544820802698550

Norman DA (1986) Cognitive Engineering. In: Norman DA, Draper SW (eds) User centered system design. Lawrence Erlbaum Associates, pp 31–61

Oehme A (2011) Ästhetisches Verständnis und ästhetische Wertschätzung von Automobildesign. Eine Frage der Expertise. Doctoral Dissertation. Universität Potsdam

Peters D, Ahmadpour N, Calvo RA (2020) Tools for wellbeing-supportive design: features, characteristics, and prototypes. Multimodal Technologies and Interaction 4(3):1–19. https://doi.org/10.3390/mti4030040

Peters D, Calvo RA, Ryan RM (2018) Designing for motivation, engagement and wellbeing in digital experience. Front Psychol 9(MAY):1–15. https://doi.org/10.3389/fpsyg.2018.00797

Pieroni M, Marques C, Campese C, Guzzo D, Mendes G, Costa J et al (2016) Transforming a traditional product offer into PSS: a practical application. Procedia CIRP 47:412–417. https://doi.org/10.1016/j.procir.2016.03.036

Pieroni de Pádua M, Marques CAN, Moraes RN, Rozenfeld H, Ometto AR (2017) PSS Design Process Models: Are They Sustainability-oriented? In: Procedia CIRP 64, S. 67–72. https://doi.org/10.1016/j.procir.2017.03.040

Placek M (2019) Market share of power tools worldwide in 2015 and 2019, with a forecast for 2027, by end users. https://www.statista.com/statistics/1228944

Pohlmeyer A, Hecht M, Blessing L (2009) User experience lifecycle model ContinUE [continuous user experience]. In: Lichtenstein A, Stößel C, Clemens C (eds) Der Mensch im Mittepunkt

technischer systeme. VDI-Verlag, pp 314–317. https://www.tu-berlin.de/fileadmin/f25/dokumente/8BWMMS/15.5-Pohlmeyer.pdf
Ries E (2011) The lean startup: how constant innovation creates radically successful. Portfolio Penguin
Rossman JR, Duerden MD (2019) Designing experiences. Columbia University Press
Rosson MB, Carroll JM (2003) Scenario-based design. In: Jacko JA, Sears A (eds), The human-computer interaction handbook: fundamentals, evolving technologies and emerging applications. Lawrence Erlbaum Associates, pp 1032–1050. https://doi.org/10.2307/798660
Sanders EB-N, Stappers JP (2014) Probes, toolkits and prototypes: three approaches to making in codesigning. CoDesign 10(1):5–14. https://doi.org/10.1080/15710882.2014.888183
Sassanelli C, Rosa P, Rocca R, Terzi S (2019) Circular economy performance assessment methods: a systematic literature review. J Clean Prod 229:440–453. https://doi.org/10.1016/j.jclepro.2019.05.019
Schmidt-Bleek F (1998) The MIPS-concept. Environmental behaviour and quality of life: ecological, sociological, psychological and communicational indicators of sustainability, 27—40
Selvefors A, Rexfelt O, Renström S, Strömberg H (2019) (2019) Use to use—a user perspective on product circularity. J Clean Prod 223:1014–1028
Sheldon KM, Elliot AJ, Kim Y, Kasser T (2001) What is satisfying about satisfying events? Testing 10 candidate psychological needs. J Pers Soc Psychol 80(2):325–339. https://doi.org/10.1037//0022-3514.80.2.325
Song W, Niu Z, Zheng P (2021) Design concept evaluation of smart product-service systems considering sustainability: An integrated method. Comput Ind Eng 159:107485. https://doi.org/10.1016/j.cie.2021.107485
Stahel WR (2019) The circular economy: a user's guide. Routledge, New York
Tuch AN, Hornbæk K (2015) Does Herzberg's notion of hygienes and motivators apply to user experience? ACM Trans Comput-Hum Interact (TOCHI) 22(4), 16:1–16:24. https://doi.org/10.1145/2724710
Tuch AN, Trusell RN, Hornbæk K (2013) Analyzing users' narratives to understand experience with interactive products. In: Proceedings of CHI 13, April 27–May 2, 2013, Paris, France, pp 2079–2088. https://doi.org/10.1145/2470654.2481285
Tuch AN, van Schaik P, Hornbæk K (2016) Leisure and work, good and bad: the role of activity domain and valence in modeling user experience. ACM Trans Comput-Hum Interact (TOCHI) 23(6):35. https://doi.org/10.1145/2994147
Tugade MM, Fredrickson BL (2004) Resilient individuals use positive emotions to bounce back from negative emotional experiences 86(2):320–333. https://doi.org/10.1037/0003-066X.60.7.678.Positive
Tukker A, Berg C, Tischner U (2017) Product-services: a specific value proposition. New Bus Old Eur, 22–34
Uhlmann J, Wölfel C, Krzywinski J (2016) Experience—a central concept in design and its roots in the history of science. In: Lloyd P, Bohemia E (eds) Future focused thinking - DRS international conference 2016, 50th anniversary conference Brighton, UK. https://doi.org/10.21606/drs.2016.365
United Nations Environment Programme (UNEP) (2022) Visualising circularity. UNEP circularity platform. https://buildingcircularity.org/
Velenturf AP, Purnell P (2021) Principles for a sustainable circular economy. Sustain Prod Consum 27:1437–1457
Vogel M, Willner M, Wölfel C, Krzywinski J (2019) User Experience Design für Sicherheitstechnik. In: Stelzer R, Krzywinski J (eds) Entwerfen Entwickeln Erleben in Produktentwicklung und Design, 2, TUD press, pp 219—230
Wallner TS, Snel S, Magnier L, Mugge R (2022) Contaminated by its prior use: strategies to design and market refurbished personal care products. Circ Econ Sust. https://doi.org/10.1007/s43615-022-00197-3

Walzberg J, Lonca G, Hanes RJ, Eberle AL, Carpenter A, Heath GA (2021) Do we need a new sustainability assessment method for the circular economy? A critical literature review. Front Sustain 1:12

Wölfel C, Krzywinski J (2019) Industrial design engineering: teaching industrial design as a specialisation within a mechanical engineering curriculum. Des Educ Today: Tech Context, Programs Best Pract, 271–292

Yoon J (2015) Positive emotional granularity cards

Zeiner KM, Burmester M, Fronemann N, Krüger AE (2017) Evaluation von Methoden zur Gestaltung positiver User Experience. In: Hess S, Fischer H (eds) Mensch und computer 2017 – usability professionals, 10.–13. September 2017, Regensburg. Gesellschaft für Informatik e.V. und die German UPA e.V., pp 1–11. https://doi.org/10.18420/muc2017-up-0155

Zeiner KM, Burmester M, Haasler K, Henschel J, Laib M, Schippert K (2018) Designing for positive user experience in work contexts – experience categories and their applications. Hum Technol 14(2):140–175. https://humantechnology.jyu.fi/archive/vol-14/issue-2/designing-for-positive-user-experience-in-work-contexts

Zeiner KM, Haasler K, Henschel J, Laib M, Burmester M (2018) Experience categories in specific contexts – creating positive experiences in smart kitchens. In: Proceedings of the international HCI conference 2018

Zink T, Geyer R (2017) Circular economy rebound. J Ind Ecol 21(3):593–602

10 Blending Design and Behavioural Science in Three Linked Public Policy Experiments Towards a Circular Economy

Stefan Kaufman and Jennifer Macklin

1 Introduction

A representative but not universal definition of Circular Economy is "an economic system that replaces the 'end-of-life' concept with reducing, alternatively reusing, recycling and recovering materials in production/distribution and consumption processes" (Kirchherr et al. 2017). However, this kind of minimalist definition renders people invisible, and neglects key aspects of sustainable development such as inter- and intra-generational equity, and broader social and biophysical outcomes articulated in the Sustainable Development Goals and the Circular Society.

As is argued in this book's introductory chapters, design approaches could be central to placing people in the centre and more successfully tackling the near-impossible challenge of decoupling the material and energy impacts of production from the human benefits of consumption in a harmonious human ecological equilibrium (Boyden 2016; O'Rourke and Lollo 2015; Parrique et al. 2019; Wiedmann et al. 2020).

Behaviour Works Australia at the Monash Sustainable Development Institute, Monash University, and Sebastian Jarvol, Australian Government Department of Climate Change, Energy, the Environment and Water.

S. Kaufman (✉) · J. Macklin
Monash University, Melbourne, Australia
e-mail: Stefan.kaufman@monash.edu

© The Author(s), under exclusive license to Springer Nature Singapore Pte Ltd. 2024

G. B. Melles and C. Wölfel (eds.), *Design for a Sustainable Circular Economy*, Design Science and Innovation, https://doi.org/10.1007/978-981-99-7532-7_10

1.1 Design and Behaviour

Working with practitioners who are trained in, and passionate advocates of, either approach, we've observed at times a level of mutual suspicion and incomprehension in early interactions. Below, we have attempted to sketch possible 'poles' of tension between the two approaches—Table 1—(after Feitsma 2018, 2019; Mori et al. 2020).

In reality, we feel the above abstractions could describe either perspective on a good or bad day. For example, some authors have argued design is itself problematic in CE. In a review of the role of people in CE design narratives and approaches, Lofthouse and Prendeville (2018) suggest that circular design typically positions people as passive subjects whose behaviour must be optimised to enhance efficiency in consumption. The authors contrast this with perspectives on design, such as human-centred design (HCD), where people are active participants attempting to realise their own narratives in the socio-technical system around them (Lofthouse and Prendeville 2018; van Dam et al. 2020). Against this, in a review of applications in global public health, HCD was found to be dominantly used to refine existing interventions rather than develop new, transformative ones (Bazzano et al. 2017), suggesting that too is a broah church with room for contradictions.

More constructively, this chapter seeks to demonstrate that both design and behaviour perspectives put people into the centre of CE (and usefully differently). Indeed, rather than exclusive alternatives, design and behavioural approaches can be complimentary. Some of the 'poles' in Table 1 may be useful to steer towards, and away, in different circumstances. For example, there are limits to agency and within that, decision-making capability; a great many day-to-day decisions will continue, by cognitive necessity, to be routinised habits (Huttunen et al. 2021; Verplanken and Wood 2006). Some level of system design, goal setting stewardship and compliance is necessary for a system to trend in a desired direction (O'Rourke and Lollo 2015).

Consequently, CE initiatives can usefully draw on both perspectives—i.e. making many sustainable behaviours the path of least resistance (McKenzie Mohr and Smith 2008) by good system design, while also recognising where transformation is needed. A middle ground perhaps is reflected in the increasing recognition that circular business models and systems of production and consumption require cooperative users to succeed (Bocken and Short 2016; Centobelli et al. 2020; Wastling et al. 2018), and that this can take place through a wide range of more active and passive roles.

This noted, we acknowledge that for better or for worse, since 2010, at least 200 governments across the world decided to create behaviour change teams—usually called 'Behavioural Insights' (BI) teams (Ruggeri et al. 2019). BI is now seen as a specific approach to the development, testing and implementation of government policy (Ruggeri et al. 2019). Sustainability is not a central focus of most government BI teams, yet behavioural insights and broader behavioural public policy are being applied in areas relevant to CE, including sustainable food consumption,

Table 1 Possible underlying tensions in the spectrum between 'behavioural insights' and 'human-centred design' approaches

	Behavioural insights		Human-centred design
Epistemological orientation	Neo-positivist, reductionist, deductive	↔	Phenomenological, holistic, inductive
Delivery style	Structured, sequential and planned	↔	Adaptive/agile, creative
Values	Libertarian, utilitarian (i.e. greatest good for the greatest number, even if this means manipulating people)	↔	Collectivist, deontological (i.e. means must be consistent with ends)
Problem definition and possible solutions are best found in…	Existing empirically validated research[a] supplemented with rigorous primary and secondary research where needed	↔	New/bespoke deep engagement with 'end users' as whole people
Experimentation is ideally	A rigorous impact evaluation testing theory in practice	↔	Rapid iteration producing outcomes that align with end users stated needs and preferences
Trust is placed in	Experts and the observations of practitioners at the coal face	↔	End users lived experience and stated needs and preferences
Certainty/ consequence appetite	High stakes, high certainty solutions show value	↔	Low certainty low stakes safe fails facilitate learning
Success is	Confidence that an intervention targets parts of important problems can usefully be tackled by someone(s) changing to a specific, expert-preferred behaviour and validated knowledge of what intervention(s) (system changes, services, programs, etc.) could be effective in changing it	↔	Improved understanding of the full ecosystem surrounding the user and what changes to design/ adapt systems, products or services appear likely to enable existing user-preferred behaviours or render system changes more acceptable

[a] Increasingly from non-academic, open-source and big data rather than traditional academic publications (Bakdash and Marusich 2022; DellaVigna and Linos 2020; Maier et al. 2022; Mertens et al. 2022; Szaszi et al. 2022)

reducing waste and encouraging resource efficiency, and firm-level environmental compliance and voluntary program participation.

1.2 BPPA—combining Behavioural and Design Insights

Rather than BI, we prefer the term "behavioural public policy and administration" (BPPA) in contrast to more narrow 'behavioural insights'—see (Kaufman et al. 2021). Here, we argued that with broadened questions asked, actors considered, lengths of time frames and the types of outcomes desired, engagement with multiple behavioural change perspectives is both deeply insightful, and can transform how government and other actors engage with policy problems. At its best, BPPA has seen policy makers and administrators working with behavioural economists, psychologists, sociologists, ethnographers, big data analysts, design thinkers and government innovation specialists in a wide range of policy areas (Ewert 2020; Feitsma 2019; Oliver 2017).

Below, we detail a series of case studies that reflect three positive trends we see emerging more broadly when governments take a BPPA approach (1) supporting recognition in policy that not only does human behaviour not always or even typically reflect rational choices based on available information, but that a wide range of internal and external factors shape behaviour, so a wide range of policy tools are needed to influence it; (2) this in turn encourages people designing policy and delivering programs and services to take a far more 'real-world', empirical and end user focused approach to public policy and administration; and (3) that this is building a norm that experimentation, user participation, research and robust measurement must be embedded into policy processes that expect to impact behaviour. Considered together, these three points make it clear that BPPA (incorporating both HCD and BI elements) has a lot to offer policy reform efforts to transition towards a sustainable CE.

In the following section, we introduce the broad context and specific program that produced the three case studies that we believe reflects a BPPA approach to advance the CE. As part of this, we briefly outline the BehaviourWorks method, a phased problem-solving approach used to plan and implement the three projects. The remainder of the chapter presents the key activities and insights from the case studies in (1) preventing household recycling contamination, (2) using consumer eco-labels to support CE outcomes and (3) encouraging business to adopt CE practices and business models. The discussion reflects on what the case studies suggest about the strengths and weaknesses of combining design and behaviour change approaches, aided by a reflection from our government partner co-author on their experience as an in-house expert and knowledge broker involved in all three projects. We conclude with implications for the role of design as both an embodiment and agent of change.

1.3 Case Study Context

Our context is Australia, and, in particular, a multi-agency collaboration across university, state and federal government that explored how behavioural approaches

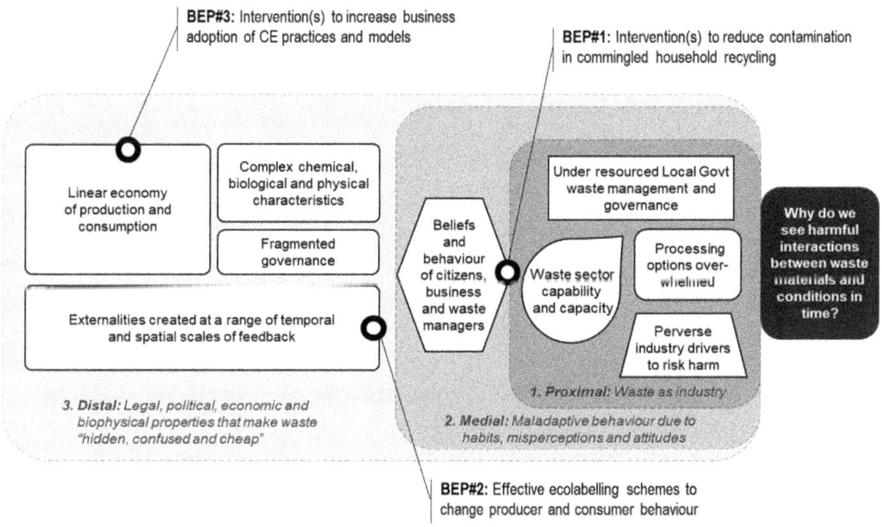

Fig. 1 Australian waste system map developed by policy makers (simplified), and selected behavioural experiments for policy

could contribute to the transition towards a CE—the Waste and CE Collaboration[1] 2019–2022 (the Collaboration). The Collaboration was conceived in 2018 within a long running partnership applying behavioural science to public policy challenges. The choice to focus on waste and CE was in direct response to the so-called 'China sword' recycling crisis, and the underlying issues it revealed to state and federal government in Australia. Investigating behavioural dimensions of CE specifically after such a disruption has been both insightful and influential, as the disruption of global low-grade recycling flows highlighted systemic dynamics and vulnerabilities in the established regime (see Fig. 1 further below).

Socio-technical disruptions in transition have been argued to increase policy makers' receptiveness to emerging niche innovations (Turnheim and Geels 2013). As the convenors of a major policy and practice workshop in 2019 observed: "The concept of the Circular Economy is not new, but rather, Australia has never had the correct mix of drivers and opportunities and these are aligning now" (Boxall et al. 2019, p. 44). Certainly, after China substantially tightened its tolerance of contamination in co-mingled recycling imports in 2018, the resulting chaos in Australia and elsewhere arguably substantially boosted engagement with CE in Australia. For example, most state government agencies began almost immediately developing and implementing CE policies (e.g. DELWP 2019; EPA NSW 2019; Green Industries SA 2020), and the federal government's National Waste Policy was updated to explicitly reference CE principles and ideas (DoEE 2018).

[1] https://www.behaviourworksaustralia.org/major-projects/waste-collaboration.

The Collaboration's key focus areas were identified in late 2018 by revisiting a pre-existing system map of issues (shown in simplified form in Fig. 1), originally created by Australia state policy makers to identify reform priorities (Clarke 2018).

The map started with the present disfunctions in waste and recycling policy, but ended up critiquing the linear economic model that existed well before the 2018–19 crisis.

At the most immediate scale of interactions to problems present in 2018, the map identifies proximal causes for disfunctions in waste and recycling in the super-category of waste as 'an industry' (as opposed to, for example, 'an essential service'). The waste and recycling industry's market-driven structure and risk appetite, and local government's limited capability and capacity to govern them, saw processing and disposal options that were inadequate and easily overwhelmed by commodity price fluctuations, let alone disruptive international trade policy. However, policy makers in 2018 also saw this Laissez-faire approach a consequence of the beliefs, attitudes and behaviour of citizens contributed that shaped the available resources and 'policy reform windows' for waste, as well as directly contributing to problems through a lack of care and interest. Underlying all this at the distal scale is a wicked mix of structural factors. The result is a range of significant harms and externalities, including for example a number of serious urban factory fires resulting from temporary stockpiling of co-mingled materials no longer exportable (EPA Victoria 2018). Air, water and land pollution, and community health and amenity impacts, as well as tax evasion, money laundering, illegal dumping and littering were also identified as unintended consequences associated with the domestic waste and recycling industry as it operated in 2019 (AELERT 2018; E&CRC 2018).

Analysing the original detailed map and working backwards from the problem supported the group of collaborating agencies to identify and agree on three areas to intervene, through behavioural experiments for policy (BEPs) (also shown above in Fig. 1). A key interest in this collaboration was whether or not behavioural public policy experiments, especially 'up the pipe' of causes in the system, could transform paradigms and mindsets, restructure information flows and change feedback loops (Angheloiu and Tennant 2020; Fischer and Riechers 2019; Meadows and Wright 2008). The BEPs therefore targeted successively upstream intervention points that could enable the transition to a CE. Our hope and intentions were that learning from BEPs would trigger policy learning and system change—especially across demand and supply side aspects—i.e. across the system.

The three BEPs were run as parallel projects, unified by similar application of specific steps of the BehaviourWorks Australia Method ('the Method') (see Fig. 2) and a common governance structure. The Method is described in detail in the *BehaviourWorks Australia Method* book.[2] In brief, it would be familiar to design thinkers as a structured approach to understanding and tackling a human-centred

[2] https://www.behaviourworksaustralia.org/the-method-book/introduction-chapter.

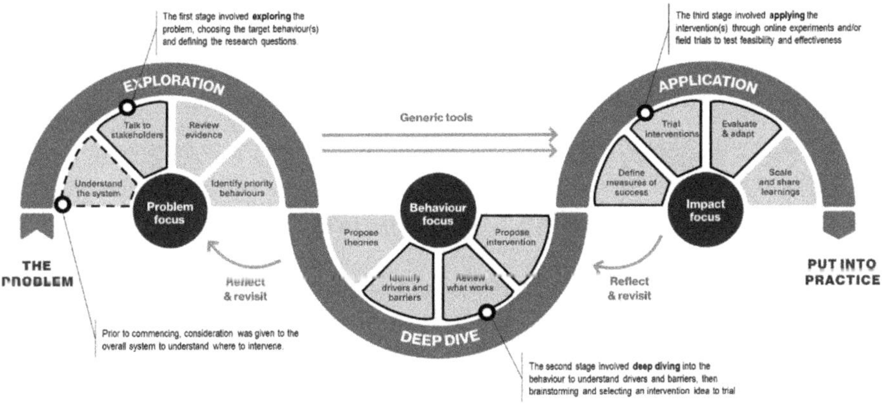

Fig. 2 The BehaviourWorks Australia Method, and selected steps incorporated in the projects

problem. Behavioural scientists would recognise it as containing the core elements of most behavioural science approaches including (1) defining the problem, (2) selecting a target behaviour, (3) identifying drivers and barriers, (4) developing intervention ideas and (5) testing intervention(s). The difference lies in the detail: different tools and methods are applied at each stage of the Method, including, as our government co-author reflected, whatever disciplinary and methodological mix helps inform a decision—(i.e. across the spectrum and poles of Table 1, p. 8).

Similarly, key stages of the Method and other behavioural science approaches have much in common with HCD processes, reflecting 'the method's' co-development with government partners using both.[3] For example, the common 'Double Diamond' process of Discover, Define, Develop, Deliver involves (1) understanding challenges and their root causes, as well as user needs and preferences, (2) define the problem to be solved, (3) develop and iteratively field-test a wide range of potential solutions and (4) build the most promising solution (see, for example, Design Council 2019; Melles et al. 2021; Victorian State Government 2020). Importantly, the Method is a menu more than prescription. This too is in common with HCD, where active User involvement is at the core of 'ideal' conceptions of HCP, but not all HCP approaches or practical applications include this. For example, the Empathic design and Contextual design HCP approaches involve the researcher moving to a greater understanding of the User but not necessarily User participation in the design process (Steen 2011).

[3] BWA has worked with a large number of government partners representing combined teams of BI, HCD and other government innovation practitioners—including co-author Sebastian Jarvol's home team. In 2021–22, we worked with the Australian Taxation Department, who have a team of over 70 staff conducting system-led design processes, and include a behavioural insights team.

The following sections provide case studies of the three BEPs, including how each incorporated both BI and HCD elements into BPPA, along with the findings and implications.

2 Case Studies

2.1 Preventing Recycling Contamination

Discovering interventions that can effectively reduce household recycling contamination behaviours was the goal of this stream. To find them, we developed a program of behavioural experiments and trials across multiple urban and regional local government areas in NSW and Victoria, the largest such trial program in Australia to date, to our knowledge.

The household contamination project implemented the key steps of the Method by[4]

- Conducting interviews with key policy stakeholders to understand the overall shape of the problem and specific contributing behaviours.
- Completing a rapid review of academic literature, along with interviews with expert 'waste educators' (local government and industry staff who educate households on correct waste behaviours) to understand the main barriers and drivers of behaviour, and what interventions have been tried before.
- Co-designing a range of potential interventions with local government implementers.
- Designing, overseeing delivery and evaluating a series of online experiments and field trials to test selected interventions.

This project had two levels of 'users': the local governments (municipal authorities) who would need to implement the final interventions/programs recommended by the Collaboration and its state government partners, and the households who would be on the receiving end of the interventions from local government.

This project adopted a more BI approach to understanding the ultimate users (households) (e.g. trusted sources in Table 1), but strongly incorporated the principles of HCD in engaging local government users in the ideation, design and testing of the interventions they themselves would need to implement. The 'end users', households, were not directly engaged in the project.

[4] Full methodological details and results can be seen in the Trial Summary Report (Downes and Kaufman 2022), available online at https://uploads-ssl.webflow.com/619ab5836de9f00d9c722d98/62e770ef2f9d22bd860e9b3a_Waste%20Collab%20Recycling%20Trials_Summary%20Report%20FINAL%20v1.1.pdf.

2.1.1 Background Research/Discovery and Definition Phases

Our background research revealed that while co-mingled recycling contamination occurs at the household consumption/disposal level, the root causes reach all the way back to the production stage, as well as all the way forward to collection, reprocessing and end-markets of recycling material (Kaufman et al. 2020b).[5] Specifically, alongside household needs, preferences, choices and behaviour, causes of contamination include the way products are designed and labelled, the way household waste is collected and processed for recycling, the availability of end-markets for recycled material and products made with recycled content, and the multiplicity of actors and conflicting messages currently involved in communicating correct waste disposal behaviour to households (Kaufman et al. 2020b).

While the China-sword crisis spurred some up- and down-stream efforts (such as standardised recycling labelling, and government-led procurement of products with recycled content), the main efforts to solve contamination currently lie with 'waste education' efforts by local government, environmental organisations and waste industry. However, while waste educators currently invest considerable time, energy and passion into initiatives to improve household waste and recycling behaviours, many feel that they are not achieving sufficient level impact on levels of contamination in household co-mingled recycling bins. A key need was therefore for evidence-based interventions that are effective in changing behaviour and reducing contamination. Such interventions needed to be feasible, regardless of the size of the local government (and their waste education budget).

While acknowledging the systemic issues, the Collaboration steering group elected to keep the problem defined at the 'behavioural' level and focus on meeting the need for evidence-informed interventions directed at households.

2.1.2 Co-design/Development Phase

Once the final problem statement and target behaviours had been defined by the Collaboration project steering committee, a co-design process was initiated. As noted above, local government practitioners are one level of 'users' in this project, and their involvement in designing potential solutions was considered essential in developing feasible intervention options. A cooperative, 'co-design' HCP approach (Steen 2011) was adopted, involving two phases. The first saw overpotential users of the intervention from local government and industry, as well as representatives of state and federal departments, coming together over two days in Sydney and Melbourne to ideate, refine and prioritise ideas for interventions. A low-fidelity prototyping approach (Burns 2018) was emulated by pre-preparing mock-ups of a whole range of potential ideas that workshop participants could draw on to brainstorm and articulate ideas, in order to enable more realistic feasibility assessments

[5] Kaufman et al. (2020b). What 'works' behaviorally to reduce contamination of recycling at the kerbside: A rapid evidence and practice review. Prepared for the BWA Waste and CE collaboration. (pp. 1–72). BehaviourWorks Australia, Monash University.

by participants. Following this, the prioritised ideas were further ranked by the Collaboration project steering group into final five types of intervention ideas, selected as valuable to the largest proportion of local government users.

The selected interventions involved a combination of BI and HCP inspired ideas. For example, the majority of ideas for the social media messages came from generic behavioural insights, while the education materials combined standard visual design principles and generic behaviour insights. However, the first two ideas for local government programs (systems and context) came from identified household pain points.

2.1.3 Trials/Delivery Approach

A total of 38 experiments/trials across three streams were initially planned, with 26 trial delivery partners across two states. Despite the Covid-19 pandemic and associated lockdowns, 22 experiments/trials with 16 delivery partners were able to be completed: 6 field trials and 16 online experiments.

2.1.4 Findings

The program of behavioural experiments identified that there are better and worse ways to intervene to reduce contamination through waste education and behaviour change programs. Key learnings on waste education effectiveness from across the three streams of experiments/trials included

- Targeting contamination (rubbish in the recycling bin) and leakage (recycling in the rubbish bin) at the same time creates confusion. **Focusing on** contaminating **'no' items is more effective at getting them out of the recycling bin** than highlighting what can be recycled ('yes' items).
- Adding behavioural messaging designed to capture attention or trigger action can increase effectiveness but it can also backfire, exacerbating the problem. (More research is still needed to better understand what types of messages are effective and when.)
- Traditional approaches (e.g. signage and educational materials) are not sufficient on their own to change behaviour. **Behaviourally informed, personalised feedback can reduce contamination.** (More research is still needed to identify other, less-intensive options, particularly for apartment buildings.)

More broadly, a known challenge with both HCD and BS projects is the scaling up of tested 'prototypes' (Hsieh et al. 2021; Mori et al. 2020). This behavioural experiment for public policy produced some important findings, which, if adopted by waste educators across local governments and the waste industry, could increase the effectiveness of current efforts, hastening change within households and associated improvements in recycling outcomes for Australia.

The speed at which the learnings become integrated into practice could be increased by the support of state and federal governments, including by actively disseminating these findings; investing in the development of common materials and programs; and providing financial support to local governments and other entities to upgrade communication materials and deliver personalised feedback programs (which can be resource intensive).

Secondly, the participatory approach taken to trial development and delivery required relinquishing of much of the control standard to BS projects. This meant that many of the findings were context-specific and therefore preliminary in nature. Only a few findings could confidently be generalisations to most Australian local governments. This means that despite the major financial investment already made in this project by the sponsors, further research is required to validate and refine our understanding of the effectiveness of the various ideas tested in the program.

Finally, as acknowledged at the beginning of the project, the problem was defined as a narrow behavioural challenge. However even if rolled out at scale, the interventions tested in the program of behavioural experiments are unlikely to completely mitigate the effects of the constantly evolving and complex packaging produced, or inconsistencies between recycling collection and processing infrastructure. Taking more of an HCP approach to problem definition may have enabled the early discovery phase to firstly determine which contamination problems can be addressed by behaviour change that aligns relatively closely to existing household context, needs and preferences, and which would be more effectively or efficiently addressed in other ways, such as through infrastructure, service or other systems change. The latter could have resulted in recommendations for current waste policy, recycling infrastructure, packaging design and product marketing that may be sending confusing or detrimental behavioural signals, resulting in unintended consequences and perverse outcomes for recycling contamination. For example, scheduling a second system mapping stage, focusing specifically on household recycling contamination, could have brought together a range of stakeholders to help identify where to prioritise household behaviour change interventions, and where to prioritise up- or down-stream changes to systems, services or products.

2.2 Eco-labels to Improve Producer and Consumer Behaviour

The goal of the second stream of work in the Waste and Circular Economy Collaboration was to understand whether and how eco-labels can be effective in increasing producer and consumer adoption of circular economy products and services. Specifically, project sponsors were interested in the potential for new labelling schemes (or expansion of existing schemes) that could support the transition to a CE. To provide a foundation for exploring effectiveness of eco-labels in Australia, BehaviourWorks investigated consumer preference for circular product characteristics, and knowledge about key CE terms and existing eco-labelling schemes. While such research is an early step on the development pathway to evidence-informed behavioural public policy, initial results indicate a promising contribution from eco-labelling schemes.

The eco-labels project implemented the key steps of the Method by[6]

- Conducting interviews with key policy stakeholders to understand the landscape of consumer eco-labels in Australia.
- Completing a rapid review of academic literature to understand the effectiveness of eco-labelling schemes around the world.
- Co-designing possible interventions with representatives of major furniture, clothing and office equipment/stationary brands.
- Designing and delivering an online survey and choice experiment to test consumer interest in circular products and whether CE labelling schemes could be effective in an Australian context.

This project had three levels of 'users': (1) administrators of future labelling schemes, (2) production and retail businesses that need to apply the label to their products, and (3) individual and business consumers who might use the label to inform their consumption choices.

Again, reflecting BI elements, participation of users limited to interviews with existing labelling scheme administers and production/retail businesses during the background research/discovery phase. More in keeping with HCD was participation by business, NGOs and regulators in the trial development and testing phases. Householder Users were not directly involved in the definition, development or delivery phases except as participants in the choice experiment and associated surveys of attitudes, values and behaviours.

2.2.1 Background Research/Discovery Phase

Our background research (Meis-Harris et al. 2021) revealed the following:

- Evidence of widespread effectiveness of labelling schemes is scarce. Voluntary eco-labelling schemes rarely reach sufficient market penetration amongst applying producers/retailers to result in widespread market shifts or changes to product portfolios. Even where eco-labels do exist, they are more likely to be successful in raising awareness of circularity characteristics and intentions to purchase (particularly amongst more environmentally conscious consumers) than translating into significant consumption changes.
- The most important element identified that increases the effectiveness of eco-label schemes amongst both producers and consumers is trust. For producers, this is about necessity, feasibility and fairness. For consumers, trust is based on transparency, accuracy and credibility. Another critical element is that the intention/objectives of the scheme are commonly valued.

---

[6] Full methodological details and results can be seen in the Trial report (Klemm and Kaufman 2020), available online at https://drive.google.com/file/d/1_zmWow0sm8Pb2E0ME24UrLh8IcAoNtra/view?usp=sharing.

- Beyond this, visible (salient) information is needed at key consumer decision points. However, effectiveness beyond a niche of producers/consumers with aligned values requires a level playing field with traditional/linear products plus some form of disruption to existing practices/habits.

Attempts to use these findings to define the 'problem' to be addressed by the following solution ideation and testing phases instead raised additional questions. A traditional behavioural science approach may have pragmatically proceeded by using generic behavioural insights to attempt to develop hypothetical labels/schemes that address the barriers and drivers identified in the international literature. Taking the more flexible design approach instead allowed the project to make the decision to revisit the Discovery phase rather than continue on.

2.2.2 Additional Research/Revisited Discovery Phase

The project proceeded to revisit the discovery phase by conducting an online survey and choice experiment to further explore the drivers/barriers of trust and common value, in the particular context in which any eco-labels would be implemented: the Australian marketplace. The purpose of this was to understand the extent to which Australian individual consumers

- value the CE characteristics that are policy-relevant to the project sponsors (recycled content, durability, repairability and recyclability)
- trust and understand existing eco-labelling schemes.

2.2.3 Findings

The online experiment found that consumers do prefer products with CE attributes when given the opportunity to choose, but price is the single most important attribute. Encouragingly, people will pay more for CE attributes, up to a point. Results indicate potential for new or expanded schemes assuring CE characteristics—in particular durability, followed by recyclability. Recycled content is also influential, but less so. The significance of price corroborates and underlines the importance of complimentary policy tools levelling the playing field, e.g. internalising the costs of virgin materials in products and reducing the cost of producing products with CE attributes.

These experimental findings suggest that effective eco-labelling schemes could help increase Australian consumer adoption of circular economy products. The experimental setting allowed us to model an ideal scenario where information on CE attributes is equally easily accessible and salient at the point of decision-making to other product information. However, real-world labels will have to overcome additional barriers to being effective—being noticed, understood and trusted.

Current understanding of CE product characteristics
Common understandings of the terms 'recycled content', 'recyclability', 'durability' and 'repairability' provide insights into how labelling schemes highlighting such attributes may be interpreted. It indicates what information labels and accompanying marketing or educational campaigns should provide.

Common misunderstandings (held by >25% respondents) around the terms 'recycled content' and 'recyclability' suggest a need for educational campaigns. Nevertheless, choice modelling results show the two concepts, properly explained, have different impacts on choice, with recyclability being the most influential. Thus, if designers wish to realise the value of designs incorporating recycled and recyclable features, they will need to communicate them effectively to people currently confused about the difference.

Similarly, when it comes to durability labelling, most consumers expect information on how long a product will stay in perfect working order. Repairability generally connotes that a professional repair firm or the company that sold it can repair, rather than consumers themselves, suggesting that discussion of 'right to repair' needs to remain broader than DIY enthusiasts.

In terms of awareness of and trust in existing eco-labelling schemes, our research shows stark differences between how well-recognised existing labelling schemes are, ranging from 9 to 93% recognition. Existing label information is moderately well-trusted. Government sourced labels are the most trusted, with private/company labels and third-party NGOs on a roughly equivalent footing. However, most consumers (59%) do not consider environmental information on products, highlighting the need for complimentary interventions.

These findings enabled the project to reach a conclusion on the definition of the problem to be solved through any eco-label intervention: how to design eco-labels to capture attention, credibly communicate key CE characteristics and reduce any intention-behaviour gaps. This provides a basis for any future behavioural public policy research on CE eco-labels. The results have also been influential in informing the recommendations of a Productivity Commission Inquiry into the 'right repair', with an invited presentation informing a recommendation to explore the development of repairability and durability labels.

A key value of implicating both user (from households) and implementer (from business and government) perspectives in this project was underlying the systemic nature of change required to realise the potential of eco-labels. It's unlikely a purely BI or HCD approach would have been able to produce such policy- and practice-relevant insights in an integrated fashion. The HCD perspective highlights the interconnected nature of the problems, while the BI approach generated highly relevant and immediately usable data and insights compatible with policy cost/benefit analysis preferred by bodies like the Productivity Commission.

2.3 Facilitating Business Adoption of CE Practices

The third stream of the Collaboration focused on gaining a better understanding of what could help business shift towards adopting CE practices and business models.

The business models project implemented the key steps of the Method by[7]

- Conducting stakeholder interviews with key policy makers to the key industries where adoption is already occurring, and opportunity is greatest.
- Conducting two rapid academic literature reviews to
  - understand characteristics of businesses and business actors, and their barriers to circular innovation
  - identify best practice around the world.
- Running an evidence-based circular workshop series (using 'Circular Strategies') to test the effectiveness of collaborative strategies.

The 'users' in this project are businesses. This project adopted a more behavioural science approach to understanding the drivers and barriers faced by businesses by reviewing previous empirical research and interviewing practitioners and experts. However, it adopted a typical human-centred design approach to ideating and developing a solution.

2.3.1 Background Research/Discovery and Definition Phases

Stakeholder interviews and a rapid review of literature (Kaufman et al. 2020a)[8] indicate that the most significant barriers in Australia for businesses deeply engaging with the transition to a CE are NOT regulation or a lack of technological capacity, but (perceived) lack of customer demand, organisational inertia and lack of collaborative capacity (see Table 2, for the full list of barriers).

2.3.2 Co-design/Development Phase

This mix of barriers lead us to define the problem we sought to answer with our trial with businesses as *How can we bring different parts of an industry 'ecosystem' together to experiment with CE practices?* Specifically, we explored how a collaborative co-design process to develop a circular initiative could itself act as an 'intervention' to reduce barriers to the adoption of CE business models.

Thanks to an existing project with colleagues at our institution, we had a strong impact case for working in fashion and textiles and some existing relationships to build upon. This foundational work highlighted a rich and divergent CE

---

[7] Full methodological details and results can be seen in the trial report (), available online at https://www.behaviourworksaustralia.org/major-project/waste-collaboration-stream-3-business-barriers-to-circular-economy-approaches and specifically: https://drive.google.com/file/d/1zIhY9ckw9BTsXvnZhtciLPvLOQgYat00/view?usp=sharing.

[8] Kaufman et al. (2020a). Business uptake of circular economy approaches: A rapid evidence review for behavioural public policy. Prepared for the BWA Waste and CE Collaboration.

Table 2 Summary of key themes in evaluative feedback re: participant experience key CE barriers

Category	Barrier	Identified before	Discussed during	Improved after
Consumer	Consumer awareness and interest	X	X	X
	Consumer willingness-to-pay	✓	X	X
	Consumer trust of CE products	X	✓	✓
Organisational	Hesitant company culture	X	✓	X
	Skills/knowledge to design & implement CE	✓	✓	✓
	Willingness to collaborate in the value chain	✓	✓	✓
Linear system	Operating in a linear system	✓	✓	✓
	Difficulty in reaching scale in Australia	✓	✓	✓
Market	Capacity to exploit CE opportunities	X	X	X
	High upfront investment costs	X	X	X
	Low virgin material prices	✓	X	X
	Examples to construct business case for adoption	X	X	✓
Regulation	Constraining regulations	X	X	X
	Circular procurement	✓	X	X
Technological	Successful examples to adapt/emulate	X	X	✓

practice experience in the Australian Textile, Clothing and Footwear (TCF) ecosystem. Some businesses were founded with CE principles; others have subsequently adopted them, while others have limited knowledge of possible CE strategies. Taking an ecosystem view meant that we included policy actors (with more or less knowledge of CE and the TCF sector), and specialised providers (e.g. recyclers, reverse logistics). This not only responded to the key barriers of willingness to collaborate across the value chain, but also provided the potential to influence both consumer demand through the influence of different actors, and internal organisational inertia by opening space for providing external legitimacy and support for 'intrapreneurs' inside participating organisations. The logic of a trial to orient the ecosystem towards CE adoption is illustrated in Fig. 4.

To guide the process, the evidence-informed Circular Strategies approach was adopted. Created by a team of Dutch researchers from Delft University of Technology (Brown et al. 2019; Konietzko et al. 2020), it aims to foster business innovation by addressing several of the 'soft systems barriers' of CE adoption in businesses. After an initial webinar pitching the process, a facilitated series of workshops with participating businesses took place. The workshops involved ideating circularity ideas, mapping collaboration, and coming up with an action plan via a pilot canvas.

2.3.3 Results

The Circularity Deck kickstarted CE discussions amongst diverse members of the textiles and fashion ecosystem. The digital format of the workshop activities increased its flexibility and reach at similar total cost to in-person. The workshops resulted in the development of an ongoing collaboration, the 'circular stories working group'[9] who produced a manual for producing circular products, using a t-shirt as a model.[10]

The workshop was also effective in building trust through the workshop activities, but not everyone could commit to experimentation. Convening across the ecosystem was valued, but there was a tension between sending sustainability specialists with CE expertise versus executives who could commit to experiments.

Evaluation of the effectiveness of the process as an intervention in itself measured impact on barriers through surveys (results are summarised in Fig. 5). The process helped attendees build skills, knowledge and networks for circular economy in their sector. It also reduced perceptions of a number of barriers, particularly some relating to the linear system and organisational capacity. However, some barriers remained unchanged, particularly some of the structural barriers, such as consumer willingness-to-pay, low prices for virgin materials and lack of regulation around circular procurement.

Like the previous two cases, this project started with a more BI flavour, seeking to understand the predictable challenges and possible solutions. However, in this case, the analysis led to choosing a largely HCD style intervention, with the main BI contribution applying a level of impact evaluation combined with process evaluation, and considering scaling implications. Again, together, both approaches achieved more than they could separately.

2.4 Discussion

Across three case studies, we see examples of a combination of HCD and BI approaches to advancing the CE through behavioural experimentation for policy. Overall, the implication is that integrating HCD and BI can play a significant role in addressing the transition to a circular economy. HCD, which focuses on understanding the needs and perspectives of users, can be used to design products, services and systems that are more likely to be used and reused, and that encourage sustainable behaviours. Behavioural science, which examines how people make decisions and take actions, can be used to understand why certain behaviours are adopted or not adopted, and to design and test interventions that encourage more sustainable behaviours. To oversimplify both—if HCD is what engineers needed to consider end users of products—behavioural public policy (incorporating both behavioural science and HCD elements) might be what technocratic policy and

[9] https://www.monash.edu/msdi/initiatives/projects/circular-economy-textiles.
[10] https://www.monash.edu/circular-fashion.

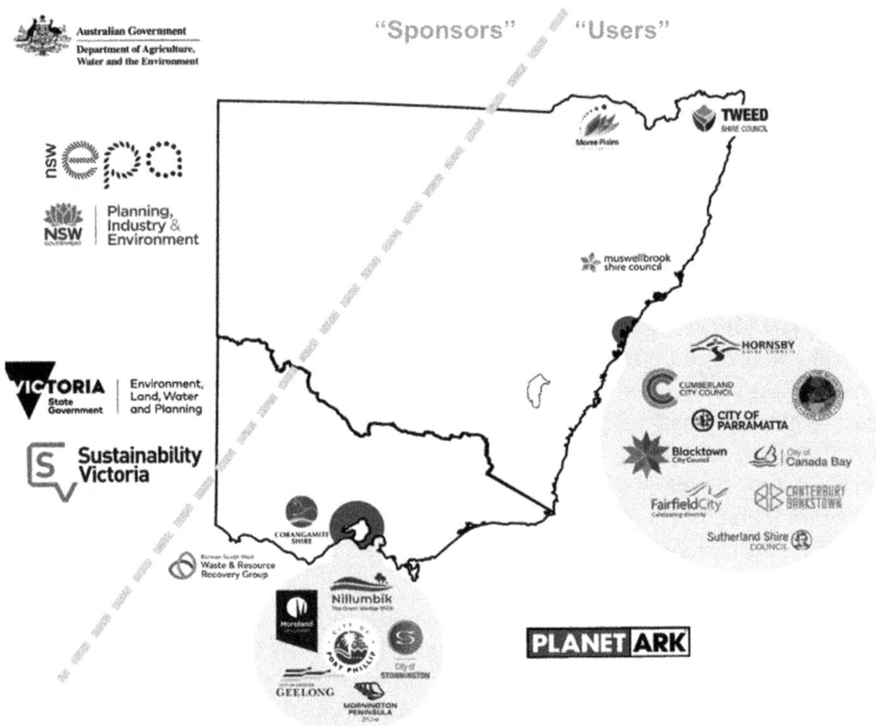

Fig. 5 Collaborating partners of the household contamination behavioural experiment for policy

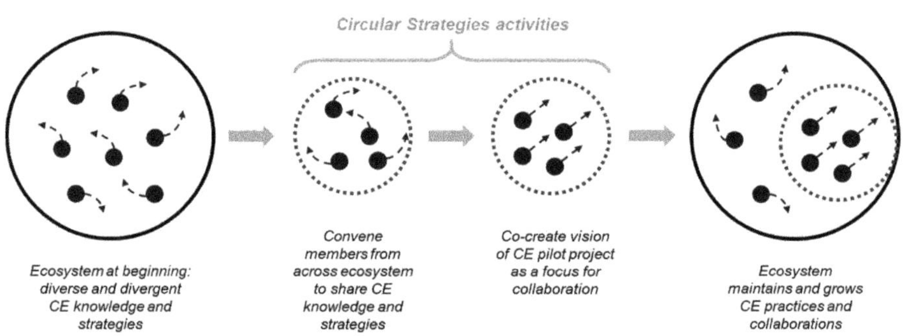

Fig. 4 CE business theory of change

program designers need to consider behavioural responses to policy and program interventions and their likely systematic consequences and dependencies.

Reflection by our government co-author (Box 1) suggests the ontological and epistemological roots of different approaches (i.e. Table 1) are less important that their relevance and contribution to decision-making at a given point in time in a given BPPA process. Underpinning it all is a fluidity of knowledge brokering provided by a mix of in-house expertise in the department, and a long-running partnership with external researchers—itself an approach to collaboration that has more in common with HCD than some BI examples readers may be familiar with.

Box 1: Integrating design and behavioural science inside government agencies
Although design (i.e. HCD above) *and behavioural economics* (BI, above) *are different fields, these approaches are often blended in an applied government context. They are not blended by doing each in series or parallel, but by ad hoc taking elements of both to suit the practical requirements of the project. Essentially both sets of tools are thrown in the toolbox, and thereafter non-experts don't ask which tool has which origin; they just pick them up and use them.*

Typically, this results in a project where, per behavioural insights, a specific target behaviour is chosen and, optionally, a method for measuring change in that behaviour is put in place. Intervention then consists of a mixture of applying common behaviour-change frameworks like EAST or the Michie wheel (Michie et al. 2014; Smith et al. 2021) *supplemented with the tools of human-centred design like ideation, qualitative empathetic interviews with the target audience and solution prototyping. There is often even further blurring with other fields/techniques like Agile, evaluation, survey methodology, and social science more broadly. This broad-church approach is necessary because applied policy contexts are, almost by definition, contested and complex, such that simple quantitative evaluation controlling for non-experimental variables is very difficult and it is often not possible or even appropriate to represent the context on a single continuum.*

(Such a) *blended approach often relies on translation by knowledge brokers to allow tools to be applied appropriately. Knowledge brokerage can be done either by a specialist team, or by enthusiastic individuals given leeway to pursue this interest as part of their usual role. However, the knowledge/experience required means policy teams usually don't maintain staff with these in addition to their subject matter knowledge requirements. Both approaches, design and behavioural economics, are still seen as novel, especially by senior leaders. This comes with the advantage of being attractive, but the disadvantage of being vulnerable to scepticism—here specialists can play a role in overcoming scepticism through internal advocacy.*

In our collaborative circular economy research, many of these features were present. Like a typical blended behavioural economics and design project, knowledge brokerage between the research consortium conducting the project and the subject matter areas who ultimately stood to benefit from the results was, though not strictly required, often very useful for increasing understanding and buy-in amongst staff in those subject matter areas. This brokerage was done by an in-house specialist team, who acted as a trusted messenger for subject matter staff, and addressed the scepticism for the unfamiliar methods which arose for a handful of these staff.

Because our collaboration was more independent and expert-led than typical purely internal projects, the methods used were much less a blend of many disparate fields than usual, and relied much more on bespoke evidence both in existing academic literature and collected specifically for the project than on generally applicable principles or frameworks. However, while the research followed a primarily behavioural economics-based method, select design methods were incorporated at various stages, notably using qualitative empathetic interviews as part of the evidence-gathering initial stage, co-design workshops in coming up with interventions and choosing an intervention whose measurement involved multiple qualitative dimensions rather than a single quantitative outcome.

Thus, although unusual in some regards, this research still represented the necessary blending of behavioural economics rigour and the complexity-embracing empathy of design which is so often required in applied government contexts.

Taken together, the three cases and our colleagues' reflection above show how BPPA creates a valuable space for integration and 'boundary work': i.e. what happens when interconnected people, places, organisations, things and processes must find a way for different perspectives, disciplines, jurisdictions and sectors to coexist (Huitema and Turnhout 2009; Van Kerkhoff 2008). This is consistent with reflections we believe would identify as BPPA in style from elsewhere (Feitsma 2018, 2019).

The individual case studies offer interesting insights about how HCD and BI can be integrated, and what benefits can be gained from doing so. For example, the eco-labels trial highlights that one way in which HCD and BI can be integrated to promote the transition to a CE is by designing products and systems that are easy to use and understand, and that align with people's values and motivations. HCD can be used to design products that are easier to disassemble and recycle, and that are made from environmentally friendly materials. Behavioural science can be used to understand the motivations behind people's purchasing decisions, and to design interventions that make sustainable products more appealing and accessible.

Similarly, there is clearly a need for designing systems and services that make it easy for people to participate in circular activities. For example, HCD can be used to design systems that facilitate the sharing and renting of products, such as a car-sharing service or a tool-sharing program. Behavioural science can be used to understand the barriers that prevent people from participating in these types of activities, and to design interventions that address those barriers.

Eco-label schemes, properly administered, well-communicated and with sufficient integrity mechanisms, can help businesses both comply with, and realise the business value of producing such items (Meis-Harris et al. 2021). However, this case study simultaneously highlights that HCD and behavioural science alone cannot fully address the transition to a CE without engaging with much wider policy tools and interventions than might be suggested by products and services. They can unpick parts of the problem, but not its entirety.

The collaborative approach taken in the CE business trial shows the value of bringing together stakeholders from different sectors and levels of society, including government, industry, community organisations and individuals, to co-create solutions and take collective action. This approach recognises that the transition to a CE is a complex and systemic challenge that cannot be addressed by any one sector or actor alone. It also acknowledges the importance of involving and engaging multiple stakeholders in the process.

More broadly still, achieving the promise of 'green growth' via CE also requires significant changes in underlying production and consumption patterns, as well as in the infrastructure and systems that support those patterns. Therefore, it's important to integrate HCD and BI with other approaches, such as systems thinking and engineering, to address the full range of issues associated with the transition to a CE. Here to though, they can support better design and application of such changes by designing policies and regulations that encourage sustainable behaviours. For example, the recycling contamination trial suggests that HCD can be used to design policies and regulations that make it easy for people to recycle and compost, such as by providing clear information about what can be recycled and where that is consistent across populations (as opposed to varied by council and waste provider, as is the case in many jurisdictions currently). Behavioural science can be used to understand why people may not recycle or compost, and to design interventions that address those barriers—in particular reducing the complexity of tasks required of committed recyclers, and banning or designing out highly problematic items such as composite packaging.

It is significant that in the preventing recycling contamination trial, we gained valuable insights into how hard it is to identify effective interventions to prevent contamination at the level of end users (households). The chosen focus on the end user in this case reflects the classic behavioural insights approach—emphasising the role of cognitive biases, social norms and decision-making processes in shaping individuals' behaviour—and seeks to design interventions that take these factors into account. Yes, post contamination feedback was shown to be effective, but focusing on providing contamination feedback as opposed to fundamental system redesign will not prevent the problem from occurring in the first place. At the same time, the project does provide a practical, fast and cheap action that can be taken in the short term to reduce contamination levels, while more ambitious system change is pursued elsewhere. A more HCD first approach may (or may not) have led to targeting a different group—for example packaging designers as the focus of the intervention (for example the Australian Recycling Label initiative arguably attacks the problem from another angle, as does the broader product stewardship program it emerged from). These however are not small changes, and require a complex array of moving parts to change to occur. It is not a bad thing to have combined efforts that complement each other in transitioning a complex system.

This noted, we do not claim to have harmoniously synthesised or integrated BI and HCD. Rather, the cases show that they can be usefully complimentary. Such tensions can be strength. A key tension between 'idealised' HCD and BI approaches (i.e. Table 1) is that BI approaches may tend to focus on changing individual behaviour in isolation, while human-centred design approaches tend to focus on changing the systems that shape behaviour in a way that users find acceptable. The respective weakness of both here for CE transitions are that while purely BI approaches might be seen as a quick-fix diversion that may not address the underlying systemic issues that lead to unsustainable behaviour, simultaneously, purely human-centred design approaches may be seen as too long term, complex and difficult to ever be realised, or place user desires in conflict with system characteristics that are not yet acceptable to change.

Similarly, another 'healthy tension' is that while behavioural insights approaches may be seen as paternalistic, as they commonly involve designing interventions that change individuals' behaviour without their active consent or involvement. Human-centred design approaches ideally involve engaging with stakeholders and co-creating solutions, which is arguably more empowering and participatory. However, the risk here is assuming that the majority of people are willing and able to participate in such engagements, make major changes and see them as desirable. The small 'n' nature of much design approaches, and the high level of involvement and investment required, may create a skewed picture of the capacity for change, as only highly motivated and engaged people generally have the capacity to participate in extensive co-design processes. The implication is that BI approaches may be most suitable when a lot of people can make a relatively small and easy change to improve the situation. But other approaches like HCD might be needed when a few individuals or groups need to come together to make a big change. Of course, CE is replete with examples of both challenges. The three cases presented here suggest we don't have to exclusively choose one or the other approach.

2.5 Conclusion

There are clear tensions in the philosophies and methods of approaches to transitioning to a CE characterised by behavioural insights (BI) and human-centred design (HCD). Behavioural Public Policy and Administration through processes like 'the Method' can be seen as an opportunity combining BI and HCD sequentially and in combination in key steps, rather than as exclusive options. While both behavioural science and human-centred design approaches have the potential to contribute to the transition towards a CE, they have different philosophies and methods. Both approaches have their own strengths and weaknesses. Some of these elements are irrelevant to users in government and practice (i.e. epistemological debates), and others are fundamental (i.e. leveraging the strengths of both approaches to generate shared problem understanding and agreement on interventions worth experimenting with).

Our hope and intention is that BPPA successfully integrates both approaches to facilitate intentional co-evolution of individual and system change. For CE, that may mean working to create long-term, sustainable change by redesigning products, services and systems to be circular by design, and in which sustainable behaviours are easy to perform. Simultaneously, we must also be able to facilitate transformative, emancipatory action amongst users, designers, governors and implementers to radically restructure parts of the system to better align with desired outcomes (i.e. where sustainable behaviours are currently very difficult or impossible). It is hard to see how we can achieve the decoupling of the outcomes that ought to motivate 'green growth'—i.e. ultimately happy, healthy human ecological systems—from the current high material and energy impact system, without doing both.

Acknowledgements This chapter would not have been possible without, and draws on reports of BehaviourWorks projects within the BehaviourWorks Australia Waste and Circular Economy Collaboration. These projects were led respectively by Jennifer Macklin (recycling contamination), Celine Klemm (Eco-labels and CE) and Alexander Saeri (business engagement with CE). The chapter also benefited from early advice from Brian Baldassare, who was also a key contributor to the business CE project.

References

AELERT (Feb 2018) Is being right enough? In: Proceedings of the annual conference. Annual conference, Sydney. https://www.aelert.net/aelert-events/2018-aelert-conference/speakers

Angheloiu C, Tennant M (2020) Urban futures: Systemic or system changing interventions? A literature review using Meadows' leverage points as analytical framework. Cities 104:102808. https://doi.org/10.1016/j.cities.2020.102808

Bakdash JZ, Marusich LR (2022) Left-truncated effects and overestimated meta-analytic means. Proc Natl Acad Sci 119(31):e2203616119. https://doi.org/10.1073/pnas.2203616119

Bazzano AN, Martin J, Hicks E, Faughnan M, Murphy L (2017) Human-centred design in global health: a scoping review of applications and contexts. PLoS ONE 12(11):e0186744. https://doi.org/10.1371/journal.pone.0186744

Bocken NMP, Short SW (2016) Towards a sufficiency-driven business model: experiences and opportunities. Environ Innov Soc Trans 18:41–61. https://doi.org/10.1016/j.eist.2015.07.010

Boxall N, King S, Kaksonen A, Bruckard W, Roberts D (2019) Waste innovation for a circular economy (Technical Report No. EP195506; CSIRO Cutting Edge Science and Engineering Symposium, p. 69). Commonwealth Science and Industry Research Organisation. https://publications.csiro.au/publications/#publication/PIcsiro:EP195506

Boyden S (2016) The bionarrative, 1st ed. ANU Press. https://doi.org/10.22459/B.08.2016

Brown P, Bocken N, Balkenende R (2019) Why do companies pursue collaborative circular oriented innovation? Sustainability (Switzerland) 11(3). https://doi.org/10.3390/su11030635

Burns C (2018). Human-centred design. In: van Gemert-Pijnen L, Kelders SM, Kip H, Sanderman R (eds) EHealth research, theory and development, 1st ed. Routledge, pp 207–227. https://doi.org/10.4324/9781315385907-10

Centobelli P, Cerchione R, Chiaroni D, Del Vecchio P, Urbinati A (2020) Designing business models in circular economy: a systematic literature review and research agenda. Bus Strateg Environ 29(4):1734–1749. https://doi.org/10.1002/bse.2466

Clarke J (2018) EPA problem solving workshop report: waste, p 71. Minds at Work on behalf of the Victorian Government Department of Land, Water, Planning and Environment, EPA Victoria and Sustainability Victoria

DellaVigna S, Linos E (2020) RCTs to scale: comprehensive evidence from two nudge units (Working Paper No. 27594). National Bureau of Economic Research. https://doi.org/10.3386/w27594

DELWP (5 Jul 2019) A circular economy for Victoria. Environment. https://www.environment.vic.gov.au/sustainability/circulareconomy

Design Council (2019) Innovation by design framework. https://www.designcouncil.org.uk/fileadmin/uploads/dc/Documents/innovation-by-design.pdf

DoEE (2018) National waste policy. Department of the Environment and Energy. http://www.environment.gov.au/

Downes J, Kaufman S (2022) Summary of behaviour change trials to reduce contamination of household recycling. Prepared for the BWA Waste and CE Collaboration. BehaviourWorks Australia, Monash University

E&CRC (2018) Waste and recycling industry in Australia (Parliamentary Inquiry ISBN 978-1-76010-776-5; Environment and Communications References Committee). Commonwealth of Australia. https://www.aph.gov.au/Parliamentary_Business/Committees/Senate/Environment_and_Communications/WasteandRecycling

EPA NSW (2019) NSW circular economy policy. Government of New South Wales, Environment Protection Authority of NSW. https://www.epa.nsw.gov.au/-/media/epa/corporate-site/resources/recycling/19p1379-circular-economy-policy-final.pdf?la=en&hash=F80151EA9C2C3E27BA889D15D18041CDF7A4D25A

EPA Victoria (2018) Combustible recyclable and waste materials guidance [Text]. https://www.epa.vic.gov.au/business-and-industry/guidelines/waste-guidance/combustible-recyclable-and-waste-materials

Ewert B (2020) Moving beyond the obsession with nudging individual behaviour: towards a broader understanding of Behavioural Public Policy. Public Policy Adm 35(3):337–360. https://doi.org/10.1177/0952076719889090

Feitsma J (2018) The behavioural state: critical observations on technocracy and psychocracy. Policy Sci 51(3):387–410. https://doi.org/10.1007/s11077-018-9325-5

Feitsma J (2019) Brokering behaviour change: the work of behavioural insights experts in government. Policy Polit 47(1):37–56. https://doi.org/10.1332/030557318X15174915040678

Fischer J, Riechers M (2019) A leverage points perspective on sustainability. People Nat 1(1):115–120. https://doi.org/10.1002/pan3.13

Green Industries SA (2020) Circular economy. https://www.greenindustries.sa.gov.au/circular-economy

Hsieh W, Slattery P, Kaufman S, Goodwin D, Saeri A, Tear M (2021) Chapter 12: making the most of an effective intervention. The BehaviourWorks Australia method book. https://doi.org/10.26180/17090318.v1

Huitema D, Turnhout E (2009) Working at the science–policy interface: a discursive analysis of boundary work at the Netherlands Environmental Assessment Agency. Environ Polit 18(4):576–594. https://doi.org/10.1080/09644010903007427

Huttunen S, Kaljonen M, Lonkila A, Rantala S, Rekola A, Paloniemi R (2021) Pluralising agency to understand behaviour change in sustainability transitions. Energy Res Soc Sci 76:102067. https://doi.org/10.1016/j.erss.2021.102067

Kaufman S, Curtis J, Saeri A, Kunstler B, Slattery P, Wild A, Bragge P, Downes J (2020a) Business uptake of circular economy approaches: a rapid evidence review for behavioural public policy. Prepared for the BWA Waste and CE Collaboration

Kaufman S, Meis-Harris J, Spanno M, Downes J (2020b) What 'works' behaviorally to reduce contamination of recycling at the kerbside: a rapid evidence and practice review. Prepared for the BWA waste and CE collaboration. BehaviourWorks Australia, Monash University, pp 1–72

Kaufman S, Saeri A, Raven R, Malekpour S, Smith L (2021) Behaviour in sustainability transitions: a mixed methods literature review. Environ Innov Soc Trans 40:586–608. https://doi.org/10.1016/j.eist.2021.10.010

Klemm C, Kaufman S (2020) Can circular economy policy leverage eco-labels? Trials research initial report. BehaviourWorks Australia, Monash University

Konietzko J, Bocken N, Hultink EJ (2020) A tool to analyze, ideate and develop circular innovation ecosystems. Sustainability 12(1), Article 1. https://doi.org/10.3390/su12010417

Lofthouse V, Prendeville S (2018) Human-centred design of products and services for the circular economy–a review. Des J 21(4):451–476. https://doi.org/10.1080/14606925.2018.1468169

Maier M, Bartoš F, Stanley TD, Shanks DR, Harris AJL, Wagenmakers E-J (2022) No evidence for nudging after adjusting for publication bias. Proc Natl Acad Sci 119(31):e2200300119. https://doi.org/10.1073/pnas.2200300119

McKenzie Mohr D, Smith W (2008) Fostering sustainable behavior: an introduction to community-based social marketing. New Society Publishers

Meadows D (1999) Leverage points: places to intervene in a system [The Sustainability Institute]. The academy for systems change. http://donellameadows.org/archives/leverage-points-places-to-intervene-in-a-system/

Meadows DH, Wright D (2008) Thinking in systems: a primer. Chelsea Green Pub, White River Junction, VA

Meis-Harris DJ, Klemm DC, Kaufman DS, Curtis DJ, Borg MK, Bragge DP (2021) What is the role of eco-labels for a circular economy? A rapid review of the literature. J Clean Prod, 127134. https://doi.org/10.1016/j.jclepro.2021.127134

Melles M, Albayrak A, Goossens R (2021) Innovating health care: key characteristics of human-centered design. Int J Qual Health Care *33*(Supplement_1):37–44. https://doi.org/10.1093/intqhc/mzaa127

Mertens S, Herberz M, Hahnel UJJ, Brosch T (2022) The effectiveness of nudging: a meta-analysis of choice architecture interventions across behavioral domains. Proc Natl Acad Sci 119(1):e2107346118. https://doi.org/10.1073/pnas.2107346118

Michie S, Atkins L, West R (2014) The behaviour change wheel: a guide to designing interventions. Silverback Publishing

Mori M, Njavika S, Fabricant R, Chhatpar R (2020) Human centred design & behavioral science integration primer [Presentation/pitch]. Dalberg Design, Dakar, Mumbai, Nairobi, London, New York. https://dig-dalberg.squarespace.com/s/Dalberg-Design-BSCIHCD-Primer.pdf

O'Rourke D, Lollo N (2015) Transforming consumption: from decoupling, to behavior change, to system changes for sustainable consumption. Annu Rev Environ Resour 40(1):233–259. https://doi.org/10.1146/annurev-environ-102014-021224

Parrique T, Barth J, Briens F, Kerschner C, Kraus-Polk A, Kuokkanen A, Spangenberg JH (2019). Decoupling debunked–evidence and arguments against green growth as a sole strategy for sustainability. EEB-The European Environmental Bureau. https://eeb.org/library/decoupling-debunked/

Planet Ark (2020) National circular economy hub. https://circulareconomyhub.org.au/

Ruggeri K, Jarke H, Berkessel J, et al (2019) Behavioral insights for public policy: concepts and cases. London, Routledge

Smith L, Slattery P, Macklin J, Kunstler B (2021) Chapter 9: using generic behaviour change tools when time and resources are scarce. The BehaviourWorks Australia method book. https://doi.org/10.26180/16544376.v1

Steen M (2011) Tensions in human-centred design. CoDesign 7(1):45–60. https://doi.org/10.1080/15710882.2011.563314

Szaszi B, Higney A, Charlton A, Gelman A, Ziano I, Aczel B, Goldstein DG, Yeager DS, Tipton E (2022) No reason to expect large and consistent effects of nudge interventions. Proc Natl Acad Sci 119(31):e2200732119. https://doi.org/10.1073/pnas.2200732119

Tamma P (21 Feb 2018) China's trash ban forces Europe to confront its waste problem. POLITICO. https://www.politico.eu/article/europe-recycling-china-trash-ban-forces-europe-to-confront-its-waste-problem/

Turnheim B, Geels FW (2013) The destabilisation of existing regimes: confronting a multi-dimensional framework with a case study of the British coal industry (1913–1967). Res Policy 42(10):1749–1767. https://doi.org/10.1016/j.respol.2013.04.009

van Dam K, Simeone L, Keskin D, Baldassarre B, Niero M, Morelli N (2020) Circular economy in industrial design research: a review. Sustainability (switzerland) 12(24):1–19. https://doi.org/10.3390/su122410279
Van Kerkhoff L (2008) Making a difference: science, action and integrated environmental research. Sense Publishers
Verplanken B, Wood W (2006) Interventions to break and create consumer habits. J Public Policy Mark 25(1):90–103. https://doi.org/10.1509/jppm.25.1.90
Victorian State Government (2020) Human centred design playbook (v1.4). Department of Premier and Cabinet, p 116. https://www.vic.gov.au/human-centred-design-playbook
Wastling T, Charnley F, Moreno M (2018) Design for circular behaviour: considering users in a circular economy. Sustainability; Basel 10(6):1743. https://doi.org/10.3390/su10061743
Wiedmann T, Lenzen M, Keyßer LT, Steinberger JK (2020) Scientists' warning on affluence. Nat Commun 11(1):3107. https://doi.org/10.1038/s41467-020-16941-y

Co-designing a Circular Society

11

Nadja Hempel, Ralph Boch, and Melanie Jaeger-Erben

The majority of political, scientific and economic measures for a transition towards Circular Economy (CE) focuses on technological and business model innovations. They largely exclude societal transformation efforts and socially innovative change attempts. Issues of improving quality of life, promoting sufficiency, changing social inequalities and unequal power relations in production and consumption systems are mainly addressed in contributions that provide a critical perspective on CE (s. e.g., Hobson and Lynch 2016; Jaeger-Erben et al. 2021; Corvellec et al. 2022). However, the Circular Economy concept has the potential to become a comprehensive social-ecological transformation program if these issues are consistently included. The term Circular Society (CS) has been introduced by a diverse range of actors from science, economy and civil society to provide a complementary or alternative framing to circular strategies. Yet the field of research and practice developing under this term is still in its infancy, and different strands of discussions have not been explored and synthesised yet.

N. Hempel (✉) · R. Boch · M. Jaeger-Erben
Hans Sauer Foundation, Munich, Germany
e-mail: n.hempel@hanssauerstiftung.de

R. Boch
e-mail: r.boch@hanssauerstiftung.de

M. Jaeger-Erben
e-mail: melanie.jaeger-erben@b-tu.de

N. Hempel
Leuphana University (formerly), Lueneburg, Germany

M. Jaeger-Erben
BTU Cottbus-Senftenberg, Cottbus, Germany

© The Author(s), under exclusive license to Springer Nature Singapore Pte Ltd. 2024

G. B. Melles and C. Wölfel (eds.), *Design for a Sustainable Circular Economy*,
Design Science and Innovation, https://doi.org/10.1007/978-981-99-7532-7_11

The Hans Sauer Foundation together with scientists from BTU Cottbus[1] and Leuphana University Lüneburg[2] initiated a co-design process[3] that is supposed to moderate the further development of the CS in terms of content, program and strategy. This process has an application-oriented focus and is intended to support initiatives in the field of the CS with target and transformation knowledge and provide political and economic decision-makers with recommendations for future actions. At the same time, a conceptual contribution to the research discourse is made by systematically summarising and structuring the conceptual foundations and further developing them discursively on the basis of several transdisciplinary co-design workshops. This chapter presents the process and its results so far and reflects upon methodological learnings and insights on key CS principles, actors and possible conflict issues. Furthermore, by working and experimenting with a combination of transformative and design-oriented research approaches, methodological and epistemological contributions to scientific practices for sustainability are made.

1 Circular Economy and the Sustainability Agenda

There is increasing awareness for the inability of global markets to respond adequately to the self-inflicted ecological damage and social injustices of ecological modernisation and climate change. Climate research as well as climate policy have acknowledged the need for a social-ecological transformation of current systems of consumption and production for decades, with the report "Limits to Growth" published by the Club of Rome (Meadows et al. 1972, 2004) and the "Agenda 21" that emerged from the United Nations Conference (1992), as two milestones in half a century of environmental debates.

A major reason for the lack of progress is seen in the linear structure of current production and consumption systems, in which nature as a source and sink of economic productivity is both overexploited and overused to create economic value (Karathanassis 2015). Due to the perceived need for a fundamental change in the resource-hungry and waste-rich linear economy, the concept of the CE has been discussed as a sustainability solution in research, business and politics for several

[1] Former TU Berlin.

[2] All findings in this chapter are based on the research conducted in the master's thesis "Roadmapping a Circular Society: Analysing and Shaping the discourse field of the Circular Society as a Starting Point for Transformative Processes" (Hempel 2022). The findings are complemented by insights gained in spring 2022, such as reflections on "Design-Driven & Transformative Research".

[3] From February 2022 to June 2023, this co-design process is funded by the Deutsche Bundesstiftung Umwelt (DBU). In the context of this project, a "Circular Society Roadmap for Germany" will be co-designed in transdisciplinary working groups, complementing the "Circular Economy Roadmap for Germany" (CEID 2021) in order to inspire and support the scientific, civil society and political sustainability landscape.

decades. CE aims to drastically reduce resource depletion as well as waste generation and is supposed to enable a decoupling of economic production and growth from resource consumption. Inherent in the CE approach is a logic of value creation that aims at value preservation rather than value consumption, shifting from a consumer to a user understanding (Geissdoerfer et al. 2017; Lazarevic and Valve, 2017; Murray et al. 2017). In addition, circular business models are said to have the potential to reduce costs, increase revenues, and provide opportunities for the financial sector (Ellen MacArthur Foundation 2020). Not only a growing number of corporations but also political decision-makers from local to international are currently pursuing more or less ambitious strategies in the direction of CE, which – as the European CE Actions Plan (European Commission 2020) shows – envisage a comprehensive re-regulation of production and distribution processes.

But while the CE concepts and strategies currently under discussion offer promising approaches towards sustainability, they have also met with considerable criticism regarding their theoretical and ideological assumptions, feasibility in practice, and social and environmental consequences (Calisto Friant et al. 2020; Corvellec et al. 2022).

A major criticism refers to the business-centrism of CE concepts and the underdetermination of societal goals. While this makes the concept easier to promote and adopt in different business contexts in the current economy, it also means that it faces inconsistencies and gaps concerning social impacts, social justice and social participation (Bauwens et al. 2020; Calisto Friant et al. 2020; Corvellec et al. 2022; Geissdoerfer et al. 2017; Hobson and Lynch 2016; Reike et al. 2018). As issues regarding its conceptualisation remain unresolved, CE has been described amongst others as a vague narrative (Niskanen et al. 2020), a horizon (Lazarevic and Valve 2017) and a "beautiful but dangerous fairy tale" (Hobson et al. 2021)[4] that only pretends to be a solution to unsustainability.

The most dominant narrative of CE in Western Countries is "circular modernism" (Bauwens et al. 2020) where CE is framed as an ecological modernisation project that focuses more on an expansion of economic activities and high-tech recovery of resources in a context of economic growth than on a fundamental rethinking of the purpose and design of economic activity itself (Bauwens et al. 2020; Hobson 2016; Korhonen et al. 2018a, 2018b; Temesgen et al. 2021). While many CE advocates, such as the Ellen MacArthur Foundation (EMF), emphasise the goal of enabling green growth through decoupling, other researchers rank the likelihood of achieving sufficient decoupling, either in a relative sense or in absolute terms, as low or clearly unrealistic (Hobson 2016; Hofmann 2022; Jackson 2016; Parrique et al. 2019) and thus plead for a post-growth (Bauwens 2021), de-growth (Schröder et al. 2019) or selective-growth circularity (Hofmann 2022). In addition to the lack of evidence that absolute decoupling can succeed, the inevitability of entropy casts doubts on the assumption that CE is feasible in a

[4] S. Kersty Hobson at CSF on "Circular Consumption" [conference presentation]; https://www.youtube.com/watch?v=95dJ6nP2jHQ&list=PLnsIHr9Ovr4Kq0OzHPSYbvij-5Tw6s883&index=15.

context of sustained economic growth (Calisto Friant et al. 2020; Korhonen et al. 2018a).

CE strategies also face limited implementation at the levels of policies, organisations and individual consumers (Corvellec et al. 2022; Kirchherr et al. 2018) and are far from achieving ecological goals, leading to rebound or counteracting effects with greater negative impacts than their linear counterparts (Bocken and Short 2016; Gonçalves Castro et al. 2022; Haupt and Hellweg 2019; Hobson and Lynch 2016; Zink and Geyer 2017; Buch et al. 2018). Likewise, social aspects are often ignored and impacted negatively (Haupt and Hellweg 2019; Hobson and Lynch 2016; Zink and Geyer 2017; Buch et al. 2018).

The focus on economic value creation and technical innovation fails to recognise the underlying, necessary and massive socio-cultural change, for instance concerning a needed change in consumption and the involvement of active citizenship. Social aspects such as global and intergenerational justice (who benefits from circular systems?), quality of life and participation in transformation are at best marginally addressed (Calisto Friant et al. 2020; Jaeger-Erben and Hofmann 2019a, b; Kirchherr et al. 2017; Kirchherr et al. 2017; Millar et al. 2019; Moreau et al. 2017; Murray et al. 2017; Pál 2022; Temesgen et al. 2019; Vanhuyse et al. 2021). This is despite practitioners seeing cultural barriers as the biggest hurdles to a shift towards more circularity (Kirchherr et al. 2017). Hence, broad participation involving active citizenship and the very heterogeneous field of actors for sustainable consumption as well as alternative forms of production, economy and cooperation is lacking. The discourse thus misses answers as to what a Circular Society might look like.

2 Circular Society: An Emerging Field in Research and Practice

In response to these limitations and criticisms, the term Circular Society (CS) has been introduced by a diverse range of actors from science, economy and civil society to provide a complementary or even alternative approach to transformations towards circular systems of consumption and production (Jaeger-Erben et al. 2021; Jaeger-Erben and Hofmann 2019a, b).

On the one hand, the CS discourse borrows from (historical) theoretical roots of CE (such as system thinking, Spaceship Earth economics and performance economy), and, on the other hand, it also makes references to debates on alternative economies (like care economy, commons, degrowth), social innovation, equity and diversity and aims to bring change to the broader community (Calisto Friant et al. 2020). While the body of literature offering a critical view of CE and crucial contributions to the discourse of CS is vast and cannot be fully captured in the above review, the body of literature that addresses CS under this term is modest (s. Table 1).

Table 1 Overview of key Contributions to the critical CE debate, and CS until 2021 (own illustration, in chronological order)

Focus	Contributors
Critical CE Debate	*Scientific literature (excerpt)*: Hobson (2016, 2019, 2020); Hobson and Lynch (2016); Geissdoerfer et al. (2017); Murray et al. (2017); Moreau et al. (2017); Kohornen et al. (2018a, 2018b); Temesgen et al. (2019); Calisto Friant et al. (2020), Bauwens et al. (2020), Pál (2022), Corvellec et al. (2022)
CS	*Scientific Literature*: Jaeger-Erben and Hofmann (2019a); Zwiers et al. (2020); Calisto Friant et al. (2020); Jaeger-Erben et al. (2021); Velenturf and Purnell (2021); Leipold et al. (2021); Melles (2021) *Grey Literature*: Jaeger-Erben and Hofmann (2019b); Boch et al. (2020a, b); HSF (2021a); Norman-Hansen et al. (2021) *Conferences, symposia and other public events*: Hofmann and Zwiers (2018); Utrecht University (2020); DBU (2020); Boch et al. (2021); Bloxhub (2021)

2.1 2018–2020

In Germany in 2019, scholars Jaeger-Erben and Hofmann (2019a, b) developed initial proposals for a sustainable and transformative CS to address the need for a "Great Transformation" (s. WBGU 2011). This was followed by a paper on Circular Literacy, proposing a knowledge framework for the transition to a CS (Zwiers et al. 2020). In the Netherlands, Calisto Friant et al. (2020) conducted an extensive literature review on CE discourses, developing a circularity discourse typology featuring a "reformative" and "transformative" CS emphasising aspects of global social justice and degrowth. In May 2020, the Second Utrecht Degrowth Symposium was held on the topic of "From Circular Economy to Circular Society" demonstrating the necessity to conceptualise a just and regenerative CS rather than an eco-efficient CE[5] (Utrecht 2020). Following up on these works, again Jaeger-Erben et al. (2021) published a perspective paper formulating core themes for a "Roadmap towards a CS".

In parallel, two foundations, the Hans Sauer Foundation (HSF) and the German Federal Environmental Foundation (Deutsche Bundesstiftung Umwelt, DBU), became active in the CS discourse field. The HSF followed up on the work of the scientists from a practical perspective and published a position paper emphasising the potential of new forms of knowledge production and participatory approaches to CS (Boch et al. 2020a, b). Shortly after, the DBU hosted a panel discussion on CS (DBU 2020) and increasingly embedded societal perspectives on CE in its funding activities.

[5] S. https://ontgroei.degrowth.net/towards-a-circular-society-an-overview-of-the-second-online-utrecht-degrowth-symposium/

2.2 2021

The initial scientific (Jaeger-Erben and Hofmann 2019a; Zwiers et al. 2020) as well as programmatic (Jaeger-Erben and Hofmann 2019b; Boch et al. 2020a, b) work on the topic led to a German transdisciplinary project consortium of the Technical University of Berlin[6] and the social design lab[7] by the Hans Sauer Foundation (HSF). Since then, the initiative aims to further develop the concept of a CS and to foster the discourse. A first milestone was the Circular Society Forum (CSF) in February 2021.[8] As a transdisciplinary conference, the forum featured international scientific and non-scientific contributions and involved more than 600 participants including key actors working on a CE, its critique (e.g., Hobson et al. 2021) and/or the idea of a CS (e.g., Calisto Friant et al. 2020). The CSF marked the beginning and the basis for the attempt to co-design CS in an open but guided process. This process and its results so far are the subject of this article.

In the following, more international scholars and practitioners from the United Kingdom, Australia, Germany, the Netherlands and Scandinavia have published on CS.

Velenturf and Purnell (2021) from the University of Leeds (UK) proposed in 2021 a value framework and a set of ten principles for the design, implementation and evaluation of a sustainable CE/CS. These proposals draw on the literature of systems ecology and the demands of sustainable development. Together with other scholars and stakeholders outside academia, they funded the Yorkshire Circular Lab, a living lab for community and knowledge building across science and practice.[9] Leipold et al. (2021) studied competing narratives of the CE in the French food sector, concluding that if CE policies are to address issues of distributive justice and equality of opportunity that often underpin wastefulness, CE narratives need to emphasise these issues. Therefore, the authors suggest further research on narratives for a CS instead of a CE, including visions of fairness. In Australia, Melles (2021) studied intermediation organisations to figure out the transition from CE to CS. In Denmark, a consortium of diverse stakeholders from academia and practice are "Designing the Irresistible Circular Society" (Norman-Hansen et al. 2021; Bloxhub 2021) as a national strategy being an official partner of the New European Bauhaus. Under the coordination of Lucia Reisch, the project "BEACON—Behavioural Insights for a Circular Society"[10] was also launched in Copenhagen to research and test behavioural changes towards a sustainable lifestyle to support building a CS.

While studies on the CE concept elaborate distinctions within CE thinking, formulate principles and develop definitions (e.g., Geissdoerfer et al. 2017; Kirchherr et al. 2017; Reike et al. 2018), the CS discourse as an alternative to established

[6] Later, BTU Cottbus-Senftenberg.

[7] S. https://socialdesign.de/.

[8] S. https://www.circularsociety.de/.

[9] S. https://spotlight.leeds.ac.uk/world-changers/circular-economy/index.html.

[10] S. https://beaconproject.dk/.

CE approaches has been little examined. Accordingly, there was a need to review the developments in recent CS discourses and at the same time to further bundle, explore, develop and consolidate the concept.

3 Co-designing Visions and Roadmaps for a Circular Society

The objective of the German collaborative roadmapping consortium is not only to study and consolidate the discourse field, but also to jointly produce action-related knowledge and new partnerships with key actors from science, civil society, business and politics. Along with strong visions – where to go – the consortium wants to foster a more coherent understanding of how to get there.

Sustainability visions can make different needs and expectations visible, create identity and community and provide an incentive for change (Vidal 2004; Wiek and Iwaniec 2014). A roadmap then takes the vision as the destination and provides a strategic plan to turn it into reality. The roadmap might reveal challenges and potentials as well as parallel pathways to overcome the first and take advantage of the second. Outlining this in a roadmap can provide a better knowledge of how different actors can contribute and participate in this transformation.

Roadmapping is commonly used to drive technology or design innovation (e.g., Phaal et al. 2004; Simonse 2017) and to guide policymaking (e.g., hydrogen roadmap for Germany, Hebling et al. 2019; CE Roadmap for Germany, CEID 2021). Often, design roadmapping processes combine the three steps of value mapping, idea mapping and pathway mapping (Simonse 2017). Although roadmapping is not popular in transformative research yet, similar approaches such as back casting and visioning (Davies et al. 2012; Quist et al. 2011; Quist 2002) are. It is assumed that transdisciplinary roadmapping holds the potential to complement visioning practices for sustainability transformation.

The roadmapping consortium is experimenting with a combination of transformative and design-oriented research approaches which integrates a varying amount of experts and practitioners from different fields of society. The intensity of participation by experts and practitioners ranges from proactive contributions and co-creative work in transdisciplinary workshops to consultation and solicitation of consent in feedback and research workshops to desk-research. Transparency, reflectivity, openness, iterative feedback loops and broad inclusion of diverse actors from different disciplines, sectors and societal levels are core elements of this process.

Table 2 illustrates the research process which is inspired by the work of Wiek and Iwaniec (2014) on visioning in sustainability science, Vidal (2004) on creative visioning conferences and Simonse (2017) on design roadmapping. The intermediate results of this process do mostly consist of protocols, answers to questionnaires and collaborative sketches and maps assembled on digital flipcharts during transdisciplinary workshops at the CSF. These results were analysed by core members

Table 2 Overview of the research process (own illustration)

Time	Research phase
Feb–Nov 21	**Framing and Initialising** Creating initial vision and roadmap material at the transdisciplinary CSF *Methods:* Conducting a digital transdisciplinary conference including participatory visioning and roadmapping workshops.[11] Applying creativity techniques such as time-travel-storytelling and writing "letters from the future,"[12] brainstorming, idea and value mapping by using the digital collaboration whiteboard "miro" *Collected Data*: – CSF contributions: Recorded sessions, discussion notes, session abstracts and contributions on the conference platform[13] – Workshop results: Text material in form of sticky notes, observation logs, participatory stakeholder maps and letters *Participation*: Over 600 participants at CSF; co-designing initial visions and roadmaps with 20–40 participants mostly from science, civil society and business sectors
	Analysis Decomposing and analysing the workshop outcomes *Method:* Inductive category building, coding and writing of memos with miro and MAXQDA *Analysed Data:* Workshop results *Participation:* desk-research and co-coding with colleagues
	Revisions, Recomposition and Finalising Revalidating, adapting and complementing the formulated categories and codes *Methods:* – Deductive revision of categories and codes in desk-research – Collaborative revision and validation in feedback sessions and research workshops[14] *Analysed Data:* – Academic and programmatic literature from the CSF discourse field – CSF contributions *Participation:* Desk-research, consultation and co-coding with CS experts from HSF and international universities
Dec 21	**Project design "Roadmap to a Circular Society – A Co-design Project for further Conceptualisation and Organisational Development of CS"** Translating the research findings into a transdisciplinary follow-up project over 1.5 years beginning February 2022. This project is funded by the DBU

[11] Documentation of the workshops: https://media2-production.mightynetworks.com/asset/21403300/CSF_Documentation_Visioning_Roadmapping_Workshops.pdf.

[12] S. https://www.sessionlab.com/methods/letter-from-the-future.

[13] Registrants were invited to introduce themselves and share their interest in CS: https://www.circularsociety.de/posts/would-you-like-to-briefly-introduce-yourself-to-the-others-and-share-why-you-are-interested-in-the-circular-society-gerne-auf-deutsch.

[14] Miro board of the research workshop: https://miro.com/app/board/o9J_llQwgkE=/?invite_link_id=611694547732.

of the roadmapping process in exchange with international CS experts and synthesised and conceptualised into a coherent CS framework. The analysis combined an "open coding" of data inspired by the Grounded Theory coding process (after Corbin and Strauss 1990) and inductive and deductive forms of interpretation (after Perry and Jensen 2001). Most data was analysed in an iterative and reflexive process which moved back and forth between empirical data, reflection of codes and conceptualisation. Also, to identify the current protagonists of CS, their interests and understandings, a descriptive actor analysis was carried out. The scope of this study is not the entire discourse on CS but is limited to empirical data from the CSF and literature published prior to February 2021. A detailed description of the participatory process and the methods used can be found in Hempel (2022).

4 Roadmapping for a Circular Society: Intermediate Results

The result of the analysis so far can be divided into three focus areas: "people", "principles" and "practices". Dividing the results into these three areas is supposed to (1) provide systematic insights into the perspectives, interests and backgrounds of CS protagonists and to particularly highlight aspects of consent and dissent amongst them (focus area "people"); (2) derive a set of core CS principles, strategies and vision themes that might provide guidance for the design, implementation, evaluation and improvement of circular projects and practices in the future (focus area "principles"); and (3) formulate recommendations on formats, process design and key topics for future transdisciplinary roadmapping activities (focus area "practices").

4.1 People: The Perspectives of Circular Society Protagonists

Most participants of the CSF and the visioning and roadmapping workshops belong to academia, business and civil society, and form a diverse group of actors from different disciplines, branches and professions (s. Fig. 1). Although most of the sectors were represented, a few to no participants joined from politics, the public sector and the social economy.

Yet, these results are based neither on a comprehensive anonymous survey nor on registration data. The sample for these results consists of responses to a public post on the forum's platform.[15] The results may be biased in that policymakers and public sector actors, as well as high-level decision-makers from the business sector, may not respond publicly to a post on the conference platform, even if they

[15] Registrants were invited to introduce themselves and share their interest in CS: https://www.circularsociety.de/posts/would-you-like-to-briefly-introduce-yourself-to-the-others-and-share-why-you-are-interested-in-the-circular-society-gerne-auf-deutsch.

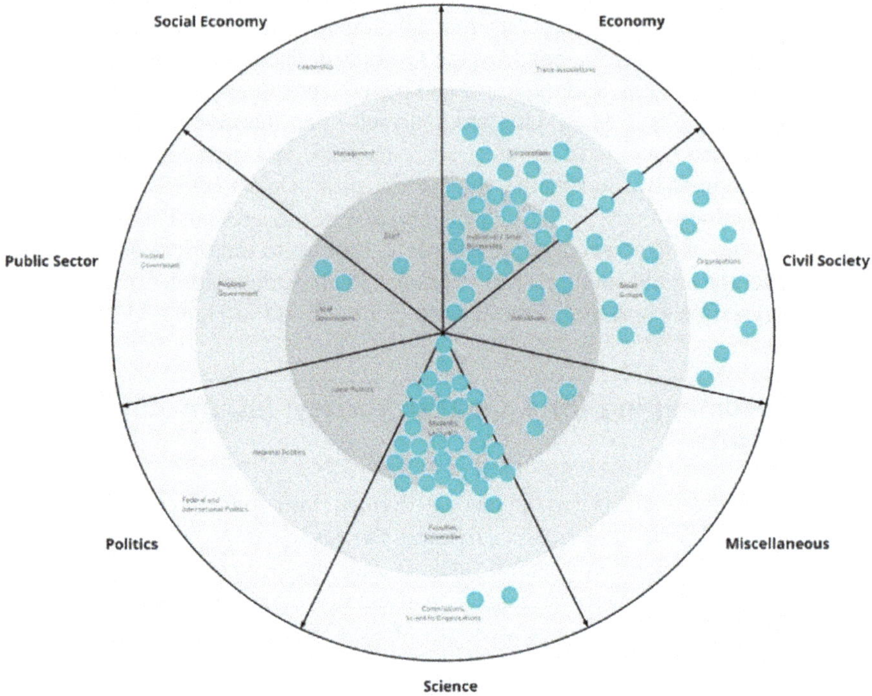

Fig. 1 Distribution of Actors participating in the CSF according to Societal Sectors and Hierarchical Levels (own illustration, based on HSF 2021b)

participated in the CSF. Anonymous enquiries at the time of registration would have been a more appropriate method for the actor's analysis.

However, the dominance of actors from academia, business and civil society over actors from the political and public sectors is considerable, and even if the methodological limitations were removed, the distribution could be different but probably not reversed.

CSF participants' interest in CS was manifold. A particular interest was in aligning CE with sustainability goals and strengthening social aspects to succeed in the transition towards a sustainable CE. Some participants aimed for CS because it is a necessary complement to existing CE approaches, and others because it differs from CE in fundamental criteria and target values. Participants also hoped that a CS conceptualisation will provide a multi-dimensional approach to operational projects. A common denominator of the proposals for a CS is that many achievements of the CE are crucial for fostering sustainability, that CE transitions are only possible with the engagement and participation of societal actors from all societal sectors, industries, levels, disciplines, etc., and that circular practices should be (re-)aligned with social-ecological goals.

The contributions of the CSF pointed out that efficiency strategies must be subordinate and embedded in consistency and/or sufficiency strategies. The major disagreement was between degrowth and sufficiency arguments on the one hand (e.g., Martin Calisto Friant[16]; Gabriela Edlinger[17]; Kersty Hobson[18]; Niko Paech[19]; Lucia Reisch[20]; Andrea Vetter[21]; Markus Wissen[22]), and green growth arguments on the other (e.g., Sonja Eser[23]; Tim Janßen[24]; Niclas Mauß[25]). The empirical data thus confirms that the question of economic growth is the "biggest elephant in the room for CE" (Calisto Friant et al. 2020). However, within the transdisciplinary workshops, sufficiency and frugality were key elements of the desirable futures discussed. Thus, a tendency for growth-criticism was apparent within CSF debates.

4.2 Principles: A Normative Framework of Circular Society

Eight CS principles, respective strategies and vision themes were inductively derived from the empirical data of the CSF and combined with deduced categories from the body of literature described above (s. Table 1). The principles illustrate core goals, values and topics of CS as well as assumed drivers and logics of transformation. Figure 2 provides an overview of the CS principles including strategies to implement those, indicated by a hashtag (#) and illustrated by sketches. An overview of the CS principles in contrast to existing approaches is provided in Table 3. The table illustrates that the CS principles are by no means completely

16 Martin Calisto Friant at CSF on "The history and plurality of circular visions" [conference presentation]; https://www.youtube.com/watch?v=BKknWZr35Ao&t=1126s.

17 Gabriela Edlinger at CSF on "Genug in einer Überflusskultur " [poetry slam], https://www.youtube.com/watch?v=eJbhbPOKkzM.

18 Kersty Hobson at CSF on "Circular consumption" [conference presentation]; https://www.youtube.com/watch?v=95dJ6nP2jHQ&t=6s.

19 Niko Paech at CSF [conference statement]; https://www.youtube.com/watch?v=Pxa-kJzvdG0.

20 Lucia Reisch at CSF [conference statement]; https://www.youtube.com/watch?v=n1rGAc9RGAE&t=356s.

21 Andrea Vetter at CSF on "Postwachstum & Kreislaufgesellschaft " [conference presentation]; https://www.youtube.com/watch?v=bIQQzJwzKbk&t=1950s.

22 Markus Wissen at CSF [conference statement], https://www.youtube.com/watch?v=qw362gmlPr4&t=1s.

23 Sonja Eser at CSF on "Thesen zur Circular Society" [conference presentation]; available under https://www.youtube.com/watch?v=lWD6zX5X2Y4&t=1460s.

24 Tim Janßen at CSF on "Umdenken für eine zirkuläre und klimapositive Welt" [conference presentation]; available under https://www.youtube.com/watch?v=lWD6zX5X2Y4&t=1460s.

25 Niclas Mauß at CSF on "Multidimensionale Nachhaltigkeitsbetrachtung zirkulärer Unternehmenstransformation "[conference presentation]; https://www.youtube.com/watch?v=lWD6zX5X2Y4&t=1462s.

new, but rather build on and complement aspects of earlier concepts. In the following, some aspects of the principles are summarised, and their potential to lever transformation is indicated.[26]

Normalise Sufficiency: A central aspect of the CSF discourse, which has been taken up in some theoretical work on the CE (e.g., Bocken et al. 2016; Reike et al. 2018) but has rarely been translated into CE application (e.g., EMF 2021), is sufficiency, i.e., forms of lifestyles and practices that are orientated towards a sufficient but not excessive consumption (as represented by CE strategies like "rethink", "refuse" and "reduce"). Embedding consistency and efficiency strategies in a general orientation towards sufficiency is understood as necessary to critical CS protagonists due to entropy, biophysical limits and the failure of decoupling economic growth and resource use. This involves questioning and rethinking understandings of prosperity and ownership as well as adapting traditional forms of work, care, policies and time that currently accelerate consumption.

Design out Waste: In cases where consumption cannot or will not be denied, there is no dispute in the CSF debate that consistency strategies are needed. As in mainstream CE concepts (e.g., EMF 2021), design is seen as having great potential to change the system from the root (s. Meadows 1999). An eco-effective, social and circular design approach has been emphasised to reduce resource demands and achieve inclusive, socially and environmentally desirable outcomes.

Keep Resources in Use: If resources are designed sustainably and circularly, there is a consensus at CSF that they should be kept in use (as conceptualised, e.g., by EMF 2021). At CSF, value creation is primarily understood as value preservation. However, there are existing stocks to manage and "changing [this existing physical structure] is rarely quick or simple" (Meadows 1999).

Build Resilience: There is consensus at the CSF that CS needs to build resilience and regenerative capacity through natural, socio-cultural and market-based diversity, regional and decentralised solutions as well as adaptability. By recognising systems as social-ecological (Becker and Jahn 2006), the needs of the ecosystem and the socio-cultural backgrounds of the actors involved are considered. A "pluriverse" was envisioned during the workshops, in which context-sensitive and diverse solutions for global goals exist.

Assure Transparency and Access: In the CSF discourse arena, the demand for transparency was complemented by the demand for equitable access to information, resources and opportunities for action ("Teilhabe"[27]). The aim is to ensure not only an efficient and effective, but also an inclusive CE. To foster sustainable consumption choices and enable participation in circular systems, open and transparent processes of value creation and destruction as well as redistribution of

---

[26] For more detail on the CS principles in contrast to existing approaches, s. Hempel (2022).

[27] While the German term "Teilhabe" is used in the meaning of access to participation, the German term "Teilnahme" refers to actual participation or the performance of participation.

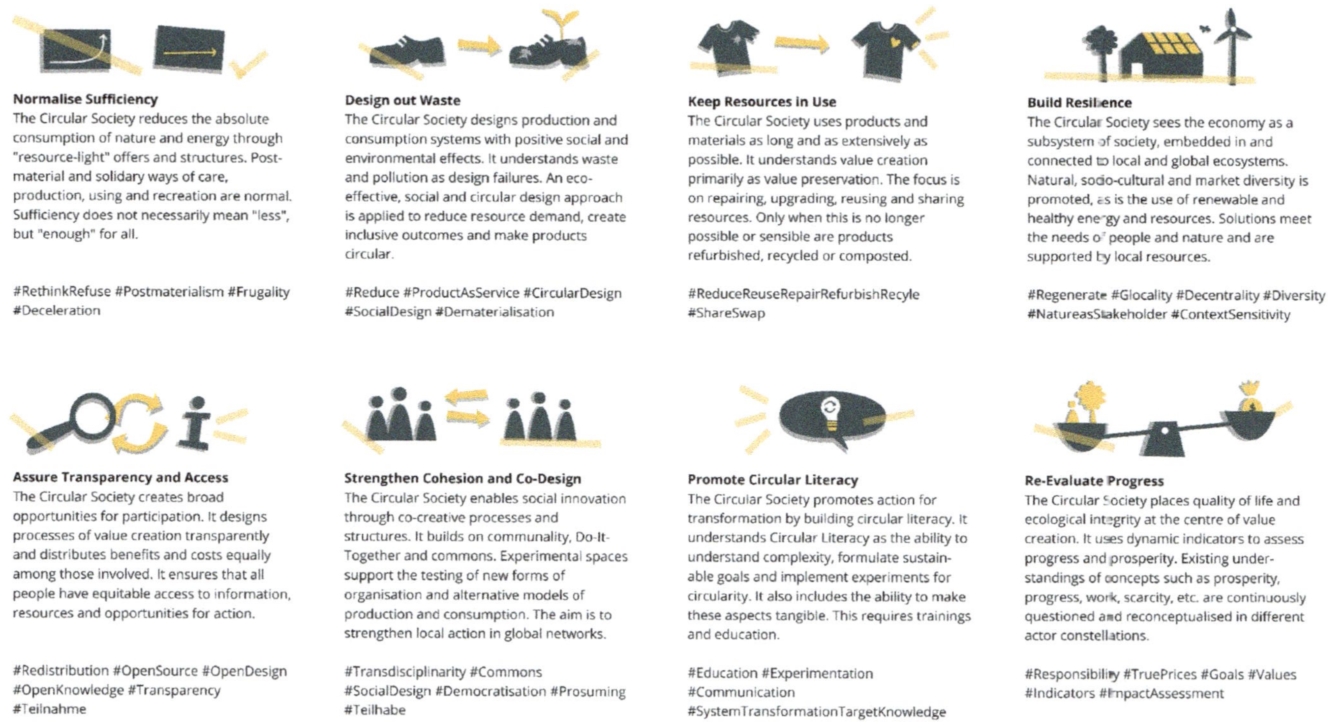

Fig. 2 CS Principles (2022 version; Illustrations: Eleonore Eisath, Hans Sauer Foundation)

Table 3 Contrasting CS Principles with existing Approaches (own illustration, based on EMF 2013; Circle Economy 2021; Bocken et al. 2016; Reike et al. 2018; Jaeger-Erben and Hofmann 2019a, b; Jaeger-Erben et al. 2021)

Approach	Principles of resource stocks and flows					Principles of underpinning system design *and* system intent			
CS Principles (based on empirical results)	P1: Normalise Sufficiency	P2: Design out Waste	P3: Keep Resources in Use		P4: Build Resilience	P5: Assure Transparency and Access	P6: Strengthen Cohesion and Co-design	P7: Promote Circular Literacy	P8: Re-evaluate Progress
CS Principles (Jaeger-Erben and Hofmann 2019a, b; Jaeger-Erben et al. 2021)	Negotiate and strengthen Sufficiency Strategies	Slowing down and closing technical and biological resource cycles				Foster democratisation, social innovativeness, social justice and solidarity			
						Accessibility and transparency	Co-creation and empowerment/ foster agency	Circular literacy	Challenge and transform capitalist value definitions
Key elements of the CE (Circle Economy 2021)		Prioritise regenerative resources	Stretch the lifetime	Stretch the lifetime Use Waste as a Resource	Prioritise regenerative resources		Collaborate for joint value creation	Strengthen and Advance Knowledge	
CE Principles (EMF 2013)		Design out waste and pollution	Keep products and materials in use	Design out waste and pollution Keep products and materials in use	Regenerate Natural Systems		Cross-cycle and cross-sector collaboration	Skills in Reverse cycle and circular product design and education	Rules of the game to quickly reach scale
10 Rs (Reike et al. 2018)	Short loops			Medium Loops Long Loops					

(continued)

Table 3 (continued)

Approach	Principles of resource stocks and flows					Principles of underpinning system design *and* system intent			
	R0 refuse	R1 reduce	R1 Reduce R2 Resell/ reuse R3 Repair	R4 Refurbish R5 Remanufacture R6 Repurpose R7 Recycle R8 Recover R9 Remine Compost Cascade					
Strategy for Resource Cycling (Bocken et al. 2016)	Narrow flows	Narrow flows	Slow flows	Slow flows Close flows	Regenerate flows				

costs and benefits were emphasised. Changing the structure of information flows and making them accessible are important leverage points for system change (s. Meadows 1999).

Strengthen Cohesion and Co-design: There is consensus within the CSF discourse arena that the establishment of processes and structures for participation and co-design ('Teilnahme'[10]) is crucial to promote (social) innovative solution development and an inclusive socio-cultural transformation. Here, the role of transdisciplinary (research) processes and heterogeneous partnerships are emphasised as well as commons. In line with Meadows (1999), changing social structures and institutions are deep leverage points for changing the mechanics of the systemic. Likewise, the empowerment of citizens to change or self-organise system structures and beliefs is a deep lever.

Promote Circular Literacy (CL): The necessary knowledge described in CE literature is limited to a systemic understanding of resource flows in the biosphere and technosphere, economic visions and technical skills. Beyond that, the concept of CL (e.g., Zwiers et al. 2020) encompasses capabilities on political and socio-cultural aspects as well as change processes. To promote CL, communication, education and experimentation strategies for circularity are demanded at CSF. Rethinking how knowledge is created, shared and used is a critical lever for sustainable transformation and can influence system parameters, feedback, design and intent (Abson et al. 2017; Meadows 1999).

Re-Evaluate Progress: While the CE literature primarily emphasises economic and secondarily ecological benefits and goals for CE, a strong understanding of sustainability is central at CSF. According to critical CS protagonists, accelerated economic growth and the resulting socio-ecological crises are to be overcome by placing social well-being and ecological integrity at the centre of desirable and resilient economic activities and policy choices (s. also Raworth 2017). This requires constant negotiation of guiding values and their metrics, consistent social-ecological value creation and assessment, as well as responsibility and accountability. With the change of core values, a paradigm shift and thus a change of the whole system is possible (Meadows 1999).

Thus, CS considers a strong sustainability approach (Döring and Ott 2001), where the economy is seen as part of society that is based on and part of a larger natural ecosystem. It aims at ecological integrity and individual and societal well-being. Well-being thereby includes the ability to participate in societal processes and to unfold identity and quality of life. The conservation of the natural life-support systems is pursued not only to safeguard human livelihood, but also as an end itself.

4.3 Practices: Roadmapping a Circular Society

The experiences of CSF revealed that methods of transdisciplinary roadmapping and visioning complement each other, are motivating and inspiring and can be introduced well in the context of a conference. However, roadmapping cannot be expected to deliver concrete milestones for achieving a complex vision in a setting of a two-hour workshop, thus longer processes are needed. For visioning, in contrast, the forum and the workshop were sufficient to gather enough initial material to be evaluated in desk work, feedback and research workshops. The following recommendations on formats, and process design might inspire future transdisciplinary roadmapping activities (in the context of CS).

Broader Involvement: To make CS relevant and applicable in different contexts, include people from the underrepresented sectors of politics, administration and citizenship in further roadmapping.

Tangible and Visual Communication: To make CS understandable and tangible to a wider audience, develop new ways and channels for simple communication.

Provocation and Prototyping: To enhance ideation and create visions, use early-stage prototyping and provocative inputs.

Creativity and Simplicity (in digital Collaboration): To engage different target groups, keep a balance between creative features and simplicity.

Consolidation: To conduct thorough visioning and roadmapping for a new discourse field, secure long-term funding.

In addition, six key topics were identified, which address in particular four CS principles that are at best marginally addressed in existing CE concepts and thus need to be further elaborated (s. Table 4).

Table 4 Key Topics for roadmapping towards a CS (own illustration, based on empirical results)

Assure transparency and access	Strengthen cohesion and collaboration	Promote circular literacy	Re-evaluate progress
Open Source and Open Design: Create structural and organisational conditions for open-source infrastructures (e.g., Open-Source Hardware Fund) and open design processes	*Collaborative Value Creation*: Experiment with organisation and business models for collaborative CS value creation *CS Hubs*: Mobilise and stabilise CS innovations locally, by making them visible and providing support	*Circular Literacy Curricula*: Develop CS curricula, both at school and university level *Communication*: Translate terms of CS into simple language, and test low-level communication channels	*CS Principles and Indicators*: Test and iterate the CS principles, sharpening the relation to other sustainability concepts. Develop indicators to measure and approach each principle

5 Core Issues to the Circular Society (Roadmap)

So far, the open and exploratory approach of the roadmapping project is suitable to capture and advance the young and dynamic CS discourse field. The mixed-data and mixed-methods approach proved adequate for analysing and shaping the diverse CSF discourse arena. Even though the process is still at an early stage, we were able to define some core issues to the CS that entail both, potentials and challenges for its further conceptualisation and implementation.

In the following, the (non)radical nature of CS and its multi-dimensional character are discussed. This is followed by an overview of the potentials and challenges for the diffusion of CS and considerations on its theoretical foundation as well as the roadmapping process.

5.1 Contested Radicality

CS has the potential to be a concept that integrates circular strategies into a framework of social, environmental and economic sustainability. Empirical evidence has shown that there is great interest in an integrated sustainability concept that thinks of CE not only at the material level but also in its social-ecological context. Within the CSF discourse arena, there is a consensus that the transition to a CE is only possible with the commitment and participation of all parts of society. Furthermore, CS protagonists agreed that a CS should be focused on environmental and social goals. In line with this, a strong sustainability approach is demanded, where the economy is seen as part of society, embedded in the natural environment. The CS principles are a concretisation of the targets and values envisioned into guidelines and strategies. They combine aspects of circular materiality with well-being, social justice, empowerment and ecological integrity. Some of them do not refer directly to the CE, but more generally to sustainability.

Sufficiency strategies, for example, do not correlate directly with circularity performance. However, if the CE is to be consistently aligned with sustainability goals, sufficiency strategies are promising. Indeed, actors involved in the CSF saw in CS the potential to link sustainability approaches that focus on consistency, efficiency and sufficiency. As Donella Meadows (1999) wrote: "[Slowing economic growth is] the same as slowing the car when you're driving too fast, rather than calling for more responsive brakes or technical advances in steering." In this sense, a motto of a CS could be "on the road to a CS, slow down the pace, find new ways of mobility and make the most of them." Thereby, sufficiency is not based on sacrifice but poses the question of the right balance, as it is common in many non-Western traditions (s. also Kallis 2019).

An obstacle for CS, however, is that the degree of radicality of transformation or reform is contested. Opinions differed on whether it is more effective to embed consistency strategies in green growth or sufficiency efforts. It could be argued that the question of how radical the transformation from linearity to circularity is represents the greatest dissent and theoretical vacuum in and between CE and CS discourses (as also analysed by Calisto Friant et al. 2020). Yet, in developing the CS principles, the normative decision was made to integrate all three sustainability dimensions. This is because the call for a sufficiency orientation was predominant at the CSF. However, in other CS discourse arenas, other foci may prevail that conflict with the CS principles and a common roadmapping process.

5.2 Momentum in the Niche?

Potential for the further development of CS and its implementation is the high resonance it receives and the current momentum for sustainability strategies. Moreover, there are pioneers whose experiences and aspirations can be built upon. Comparing the results of the CSF with the recent CE debates, it is evident that science and civil society are currently more represented in the CS discourse field. At the CSF, especially young academics pushed for a social-ecological and transformative CS. The response to the CSF with over 600 participants and numerous contributions is one example of the existing interest in the topic.

In general, due to increasing pressure for sustainability solutions, the momentum for sustainability strategies is increasing at different levels and in different sectors. CE is seen as an important building block for sustainability in this context but is coming under criticism in its conception. Criticism comes not only from edge disciplines of science, but also from established science academies and science advisory councils for policy, which are calling for an expansion of the CE debate to include social aspects and a sufficiency orientation (s. EASAC 2016; WGBU 2020). The high resonance of CS in the scientific community and civil society holds the potential to theoretically underpin and practically test circular practices, currently driven by the private sector.

On the other hand, CS has not even emerged into an established niche and faces the challenge of raising its political and public profile without weakening its transformative character. Although perspectives and approaches of non-academic origin were included in the empirical material of the conference, most literature and empirical data in this study come from academia. While CE agendas are gaining momentum in (inter)national and local politics, in the private sector and in general academia beyond the niche of transformative research, the visions of CS might be disregarded as it would require societal and economic restructuring (s. Calisto Friant et al. 2020). For example, scientific institutions and governmental funding agendas still mainly focus on technical CE approaches. A connected challenge is to find a tangible and understandable communication of CS that reaches different target groups without becoming superficial. While the usual CE visualisations have been criticised, the development of an alternative graphic is still pending.

5.3 Theoretical Scope

The CS principles have been developed in line with and in distinction to prominent CE principles and thus build on CE experiences. They also incorporate various theoretical and conceptual ideas from other concepts and projects that consider complex socio-cultural contexts of the CE and understand social, economic and environmental systems as embedded. Rather than simplifying the CE discourse, CS complements the principles of stocks and flows with principles of system intent and design, which, according to Meadows (1999), have the greater leverage for transformation.

This work does not aim to introduce a new concept with CS, but to highlight the pitfalls of prominent CE definitions and align them with strong sustainability goals. The integrated perspective on environment and society breaks down the dualism between the two and continues the tradition of CE within other scientific discourses such as social ecology, posthumanism and post-anthropocentrism. In addition, the discussions have been enriched by numerous perspectives from other scientific and cultural concepts, e.g., degrowth, post-growth, commons, sharing economy, postcolonialism and feminist economics. This provides a first, albeit immature, basis for further theoretical conceptualisation in this field.

On the other hand, the transformation narratives, CS principles and the design-based approach of experimental implementation of alternative models within CS need to be reviewed, revised and extended in terms of their socio-scientific grounding, plausibility and effectiveness. Revisions should also be made where CS criticises CE but fails to provide answers itself. For example, challenges of CE, such as the nexus between energy and biodiversity, are not solved in the discourse field of CS. Though it is often claimed that global considerations are crucial for a CS, the concept mainly refers to the living conditions of the Global North. The perfect and thus misleading image of the circle as a motif of CE is criticised in the CSF discourse arena, without being contrasted with any other proposal yet. Furthermore, discussions on a deeper theoretical foundation of the principles are necessary.[28] Another challenge is that transformation narratives and experimental implementation of alternative models should be tested for their sociological grounding and effectiveness (Blühdorn et al. 2018). Particularly in the case of visioning, the narratives partly aimed at romanticising practices of pre-modernity and rely on either the mere empowerment of local actors or a strong state. Power relations, constraints and structures of unsustainability are only marginally addressed.

[28] For example, with findings from the degrowth movement (e.g., Latouche), feminist theory (e.g., Winkler, Hofmeister), alternative economies and value theories (e.g., Marx; Sen & Nußbaum), posthumanism (e.g., Haraway, Barad, Bennet, Tsing) and complex systems theories (e.g., Bateson, Meadows); as well as with the sustainable CE principles by Velenturf and Purnell (2021).

5.4 Design-Driven and Transformative Research

In recent years, the increasing awareness of global challenges has given rise to scepticism about "traditional" understandings of science and increasingly promotes new forms of socially and ecologically engaged knowledge production on the level of disciplines, institutions and the science system as a whole (Schneider et al. 2021). Within this discourse in science, politics and society, transformative transdisciplinary forms of research, combining science, practice and engaged agendas are attributed to great potentials (Lawrence et al. 2022). In practice, scientific knowledge production and the collaborative and participatory development of sustainable futures still have to bridge many methodological and theoretical gaps, ranging from the initial framing of projects and joint knowledge production and integration to the observation and measurement of their impacts (Bergmann et al. 2021; Steelman et al. 2021).

Design-driven approaches with their emphasis on knowledge and research-based action, creative work, facilitation of multi-actor processes, prototyping and implementing could be a valuable ingredient and addition to transdisciplinary research (Gonera and Papst 2019; Franklin 2022). In the context of transdisciplinary research processes, design only is of interest, when understood as a transformative practice that "devises courses of action aimed at changing existing situations into preferred ones" (Simon, 1969).

In business and innovation contexts, design (thinking) approaches are already ascribed generic potentials in addressing complex challenges, when interdisciplinary teams are guided through collaborative and iterative processes of understanding, ideating and testing (Carlsgen et al. 2016), and are also widely used in the context of circular economy innovations (Santa-Maria et al. 2022). Design approaches also have been ascribed "great potential for promoting the systemic change, creativity, collaboration, empathy, and empowerment that are necessary for a sustainability shift." (Bertella et al. 2021). Design in general and even more so design approaches like participatory design and social design offer methods to collaborative knowledge production and mutual and social learning processes (Bijl-Brouwer et al. 2021). Design methods with their "user-centred", egalitarian, low threshold, activating and generative approach can expand the methodological canon of working in heterogeneous teams (Peukert and Vilsmeier 2019; Peukert and Vilsmeier 2021).

Also in view of the possible results, this combination of science and design seems to be promising and purposeful. The generation of system, transformation and target knowledge is commonly considered to be central to transdisciplinary research (Hirsch Hadorn et al. 2008). One strength of disciplinary science certainly lies on the side of systems knowledge, whereas design approaches can be very useful and productive additions to the generation of transformation and target knowledge.

Design also has competencies to develop processes that link divergent and convergent phases to a whole, combining situations of openness with moments of synthesis and negotiating "minimum viable" – other than scientific – results that are the basis for further collaborative efforts and an iteration-oriented continuation. This can support the explanatory and open character of transdisciplinary projects when understood as collective search processes for sustainable futures (Deserti et al. 2022).

However, design as an academic discipline and as a profession only deals with processes of social change or transformation in niches and emerging fields such as transformational design (Prendeville and Koria 2022). But even then, design theory and practice rarely refer to knowledge from disciplines like sociology or political sciences, to theories of social change or current research on transformation processes (Jonas et al. 2015; Schönfeld 2020). As a profession, design is strongly coupled to commercial and market contexts and its methods and concepts have to be adapted to the context of transdisciplinary research (Nelson and Stolterman 2012). Some of the above-mentioned potentials of design are connected to epistemological and structural weaknesses, namely its focus on quick (product and service) fixes, on fast idea generation, and its manifold affirmative relations to the current system of consumer society with its fast product-innovation cycles (Seitz 2020). Another methodological problem to solve is the quality, scope and the evidence-base of reflection and adaptation moments. As much as design has the ability to foster processes of co-producing knowledge and mutual learning, it does not offer (as most other disciplines) answers to conceptual and strategic questions of transformation. Long-term processes of change and deep-rooted structures, routines and values for example on the individual, social, cultural and political levels are outside the focus of design as much as power relations in modern capitalistic societies and path-dependencies of socio-technical systems are (van der Bijl-Brouwer and Malcolm 2020).

The integration of design approaches adds to a general challenge of transdisciplinary processes: The hybridisation of knowledge production brings about problems of legitimation and acceptance. The goals of design include not so much the production of scientific and valid knowledge, but rather the development of feasible solutions that can be iterated in testing. This poses challenges for the scholars involved, as this knowledge production will encounter problems in meeting widespread academic standards and norms.

6 Implications and Outlook

The CSF has helped to bring together actors involved in CS in practice or theory, thus focusing and further developing the debate. It can be seen as a catalyst for building the field. However, to unfold the CS concept's potential to establish a multi-dimensional approach to a sustainable CE and to motivate action, conceptual work is necessary. The CS principles are to be further developed, adapted and tested in practice. Thereby, the CS principles need to be underpinned by empirical examples, theoretical models and pilot projects. This revision should take place in different contexts within and outside the CS discourse field under study.

At the same time, the CS principles should be prevented from being misused for green- and blue-washing. Therefore, it is proposed to develop indicators for each principle to measure success at all levels and to concretise promising strategies for different sectors such as businesses or cities. However, the proposed principles are not a blueprint for sustainable action but need to be complemented by other approaches and adapted to specific contexts.

Furthermore, it is the task of the scientific CS community to improve science policy and science-society interfaces through tangible communication and a new visualisation that does justice to the multi-dimensional approach of a CS. Also, while sufficiency strategies are emphasised within the CS visions, no suggestions for the next steps towards such were made within the roadmapping workshop. Yet, the other recommendations on roadmapping towards a CS can guide future research on the CS. Therefore, long-term funding opportunities for transdisciplinary CS research projects are required.

While the strengthening of the academic sector can lead to a consolidation of the theoretical claims of the CS debate, greater involvement of political and economic decision-makers is needed to consolidate and implement the concept in practice. This also applies to civil society, which was represented at the CSF but mainly through organised initiatives. The systems, target and transformation knowledge produced offer a foundation for future roadmapping projects towards a CS. Here, the CS principles and vision themes have the potential to support CS ideation and literacy processes. Practitioners from different nationalities and societal sectors are invited to iterate on the CS principles, testing and refining them in their specific contexts such as companies, public institutions and non-profit organisations. This way, the CS principles can provide guidance for the design, implementation, evaluation and improvement of circular projects and practices in the future.

What's next? By inviting around 40 experts to participate in the "Roadmap to a Circular Society" co-design process funded by the German Federal Environmental Foundation, the aim is to further develop previous efforts such as this study's results at programmatic, strategic and content levels. Since April 2022, the consortium has facilitated the first phase of the 1.5 years roadmapping process including workgroup meetings. The key topics of the working groups are in accordance with the findings of this study: Circular Citizens and Communities, Open Source and Open Design, Collaborative Value Creation and Circular Literacy. The necessary ideas, knowledge, skills and experiences from across the German-based discourse field shall be brought together in the process to build capacity and provide space for new partnerships and projects. It is hoped that intense working sessions, exchange and discussions amongst this group can result in a compelling roadmap that resonates across society, economy and politics and sets a direction for action as well as for needed conditions. The process seeks to break new ground between transformative and design-driven (research) approaches with an open, iterative and creative character. The results are therefore open and to be defined amongst the participants, yet they shall entail further development of conceptual foundations, good practice collections and a (policy) paper that provides a better knowledge of how key actors can contribute and participate in this transformation.

References

Abson DJ, Fischer J, Leventon J, Newig J, Schomerus T, Vilsmaier U, Von Wehrden H, Abernethy P, Ives CD, Jager NW, Lang DJ (2017) Leverage points for sustainability transformation. Ambio 46:30–39. https://doi.org/10.1007/s13280-016-0800-y

Bauwens T, Hekkert M, Kirchherr J (2020) Circular futures: what will they look like? Ecol Econ 175:106703. https://doi.org/10.1016/j.ecolecon.2020.106703

Bauwens T (2021) Are the circular economy and economic growth compatible? A case for post-growth circularity. Resour Conserv Recycl 175. https://doi.org/10.1016/j.resconrec.2021.105852

Becker E, Jahn T (2006) Soziale Ökologie: Grundzüge einer Wissenschaft von den gesellschaftlichen Naturverhältnissen. Campus Verlag, New York/Frankfurt

Bergmann M, Schäpke N, Marg O, Stelzer F, Lang DJ, Bossert M … Sußmann N (2021) Transdisciplinary sustainability research in real-world labs: success factors and methods for change. Sustain Sci, 1–24. https://doi.org/10.1007/s11625-020-00886-8

Bertella G, Lupini S, Romanelli CR, Font X (2021) Workshop methodology design: innovation-oriented participatory processes for sustainability. Ann Tour Res 89.https://doi.org/10.1016/j.annals.2021.103251

Bijl-Brouwer, van der M, Kligyte G, Key T (2021) A Co-evolutionary, transdisciplinary approach to innovation in complex contexts: improving university well-being, a case study. She Ji: J Des, Econ, Innov 7(4):565–588.https://doi.org/10.1016/j.sheji.2021.10.004

Blühdorn I, Butzlaff F, Deflorian M, Hausknost D (2018) Transformationsnarrativ und Verantwortlichkeit. IGN-Position Paper, Institut für Gesellschaftswandel und Nachhaltigkeit, Wirtschaftsuniversität Wien, Die gesellschaftstheoretische Lücke der Transformationsforschung

Boch R, Gallen J, Hempel N (2020) Wege zu einer Circular Society. Potenziale des Social Design für gesellschaftliche Transformation. Hans Sauer Foundation. https://socialdesign.de/wp-content/uploads/2020/04/200420_HSS_Paper_CircularSociety_online.pdf

Boch R, Gallen J, Hempel N, Hofmann F, Jaeger-Erben M, Lersch B, Zwiers J (2020) Circular society gathering 2020 [digital gathering]. Hans Sauer Foundation, TU Berlin

Boch R, Gallen J, Hempel N, Hofmann F, Jaeger-Erben M, Lersch B, Oppelt M, Sachtleben J, Schlegel A, Trawinski C, Zwiers J (2021) Circular society forum 2021 [digital conference]. Hans Sauer Foundation, TU Berlin.www.circularsociety.de

Bloxhub (2021) Designing the Irresistible Circular Society [science talks]. https://www.youtube.com/playlist?list=Plab32e77XDC-pPIQhjcgZt2aV_SdRiY52

Bocken NM, De Pauw I, Bakker C, Van Der Grinten B (2016) Product design and business model strategies for a circular economy. J Ind Prod Eng 33(5):308–320. https://doi.org/10.1016/j.eist.2015.07.010

Bocken NM, Short SW (2016) Towards a sufficiency-driven business model: experiences and opportunities. Environ Innov Soc Trans 18:41–61. https://doi.org/10.1016/j.eist.2015.07.010

Buch R, O'Neill D, Lubenow, C., DeFilippis, M. & Dalrymple M (2018) Collaboration for regional sustainable circular economy innovation. In: Marques J (ed) Handbook of engaged sustainability (Bd. 3, S. 703–728). Springer International Publishing. https://doi.org/10.1007/978-3-319-71312-0_24

Calisto Friant M, Vermeulen WJV, Salomone R (2020) A typology of circular economy discourses: navigating the diverse visions of a contested paradigm. Resour Conserv Recycl 161:1–19. https://doi.org/10.1016/j.resconrec.2020.104917

Carlgren L, Rauth I, Elmquist M (2016) Framing design thinking: the concept in idea and enactment. Creativity and Innovation Management 25(1):38–57. https://doi.org/10.1111/caim.12153
Circle Economy (2021) Key elements of the circular economy.https://www.circle-economy.com/resources/the-key-elements-of-the-circular-economy-framework
Circular Economy Initiative Deutschland (CEID) (2021) Circular Economy Roadmap for Germany. Acatech/SYSTEMIQ: Munich/London 2021. https://static1.squarespace.com/static/5b52037e4611a0606973bc79/t/61482d35b9cffc5bb5816d1e/1632120122091/Circular+Economy+Roadmap+for+Germany_EN.pdf
Corbin JM, Strauss A (1990) Grounded theory research: procedures, canons, and evaluative criteria. Qual Sociol 13(1):3–21
Corvellec H, Stowell AF, Johansson N (2022) Critiques of the circular economy. J Ind Ecol 26(2):421–432. https://doi.org/10.1111/jiec.13187
Davies AR, Doyle R, Pape J (2012) Future visioning for sustainable household practices: spaces for sustainability learning? Area 44(1):54–60. https://doi.org/10.1111/j.1475-4762.2011.01054.x
Deserti, A., Real, M., & Schmittinger, F. (2022). Co-creation for Responsible Research and Innovation: Experimenting with Design Methods and Tools. Springer Nature.
Deutsche Bundesstiftung Umwelt (DBU) (2020) Von der Circular Economy zur Circular Society? Wege zu einer echten Kreislaufwirtschaft [digital symposium]. DBUdigital Symposium zum DBU Deutschen Umweltpreis. https://www.dbu.de/550artikel38765_2441.html
Döring O, Ott K (2001) Nachhaltigkeitskonzepte. Zeitschrift für Wirschafts-und Unternehmensethik 2(3)
Ellen MacArthur Foundation (EMF) (2013) Towards the Circular Economy. Economic and business rationale for an accelerated transition.https://ellenmacarthurfoundation.org/towards-the-circular-economy-vol-1-an-economic-and-business-rationale-for-an
EMF (2020) Financing the circular economy: Capturing the opportunity.https://www.ellenmacarthurfoundation.org/publications/financing-the-circular-economy-capturing-the-opportunity
EMF (2021, August 02) Circular economy introduction. https://ellenmacarthurfoundation.org/topics/circular-economy-introduction/overview
European Academies Science Advisory Council (EASAC) (2016) Indictors for a circular economy. Halle.https://easac.eu/fileadmin/PDF_s/reports_statements/Circular_Economy/EASAC_Indicators_web_complete.pdf
European Commission (2020) A new circular economy action plan: for a cleaner and more competitive Europe. EC: Brüssel
Franklin A (2022) Co-creativity and engaged scholarship: transformative methods in social sustainability research. Springer Nature
Geissdoerfer M, Savaget P, Bocken NMP, Hultink EJ (2017) The circular economy – a new sustainability paradigm? J Clean Prod 143:757–768. https://doi.org/10.1016/j.jclepro.2016.12.048
Gonera, A., & Pabst, R. (2019). The use of design thinking in transdisciplinary research and innovation consortia: challenges, enablers, and benefits. J Innov Manag 7(3):96–122. https://doi.org/10.24840/2183-0606_007.003_0006
Goncalves Castro C, Hofmann Trevisan A, Pigosso DA, Mascarenhas J (2022) The rebound effect of circular economy: definitions, mechanisms and a research agenda. J Clean Prod, 131136 https://doi.org/10.1016/j.jclepro.2022.131136
Hadorn GH, Biber-Klemm S, Grossenbacher-Mansuy W, Hoffmann-Riem H, Joye D, Pohl C, … , Zemp E (2008) The emergence of transdisciplinarity as a form of research. In: Handbook of transdisciplinary research. Springer, Dordrecht, pp 19–39
Hans Sauer Foundation (HSF) (2021a) Orte zirkulärer Praxis. Wirkungsmatrix. https://www.hanssauerstiftung.de/inhalt/uploads/OZP_CaseStudy_digital.pdf
Hans Sauer Foundation (HSF) (2021b) Die multi-stakeholder-system-map. social design prozesse vorbereiten. https://socialdesign.de/portfolio/multi-stakeholder-system-map/
Haupt M, Hellweg S (2019) Measuring the environmental sustainability of a circular economy. Environ Sustain Indic 1:100005. https://doi.org/10.1016/j.indic.2019.100005

Hebling C, Ragwitz M, Fleiter T, Groos U, Härle D, Held A, … , Wietschel M (2019) Eine Wasserstoff-Roadmap für Deutschland. Fraunhofer Institut für System-und Innovationsforschung ISI & Fraunhofer-Institut für Solare Energiesysteme

Hempel N (2022) Roadmapping a Circular Society: analysing and shaping the discourse field of the Circular Society as a starting point for transformative processes. [Master's thesis, Leuphana University Lüneburg]. Socialdesign Ressourcen.https://socialdesign.de/portfolio/roadmapping-a-circular-society_hempel/

Hobson K (2016) Closing the loop or squaring the circle? Locating generative spaces for the circular economy. Prog Hum Geogr 40(1):88–104. https://doi.org/10.1177/0309132514566342

Hobson K (2019) Small stories of closing loops': social circularity and the everyday circular economy. Clim Change 163(1):99–116. https://doi.org/10.1007/s10584-019-0248

Hobson K (2020) The limits of the loops: critical environmental politics and the Circular Economy. Environ Polit, 1–19 https://doi.org/10.1080/09644016.2020.1816052

Hobson K, Lynch N (2016) Diversifying and de-growing the circular economy: radical social transformation in a resource-scarce world. Futures 82:15–25. https://doi.org/10.1016/j.futures.2016.05.012

Hobson K, Holmes H, Welch D, Wheeler K, Wieser H (2021) Consumption Work in the circular economy: A research agenda. J Cleaner Prod 321:128969

Hofmann F, Zwiers J (2018) Circular society. Eine pluralistische und emanzipatorische Alternative zur Circular Economy [lecture]. Bits & Bäume, Berlin.https://media.ccc.de/v/bub2018-207-circular_society

Hofmann F (2022) Circular economy and economic (de-)growth? Let's shift the baselines! Resour, Conserv Recycl 187:106604. https://doi.org/10.1016/j.resconrec.2022.106604

Jackson T (2017) Prosperity without growth: foundations for the economy of tomorrow

Jaeger-Erben, M. & Hofmann, F. (2019a). *Kreislaufwirtschaft – ein Ausweg aus der sozial-ökologischen Krise? Schriftenreihe Nachhaltigkeit: Bd. 5.* Hessische Landeszentrale für politische Bildung.

Jaeger-Erben M, Hofmann F (2019b) From take-make-dispose to a circular society. Introduction of a new vision in six propositions. TU Berlin, Research Group Challenge Obsolenscence

Jaeger-Erben M, Jensen C, Hofmann F, Zwiers J (2021) There is no sustainable circular economy without a circular society. Resour Conserv Recycl 168:105476

Jonas W, Zerwas S, von Anshelm K (2015) Transformation design. Perspectives on a new design attitude. Birkhäuser, Basel

Kallis, G. (2019). *Limits: Why Malthus was wrong and why environmentalists should care.* Stanford University Press.

Karathanassis A (2015) Kapitalistische Naturverhältnisse. Ursachen von Naturzerstörungen–Begründungen einer Postwachstumsökonomie. Hamburg

Kirchherr J, Reike D, Hekkert M (2017) Conceptualizing the circular economy: an analysis of 114 definitions. Resour Conserv Recycl 127:221–232. https://doi.org/10.1016/j.resconrec.2017.09.005

Kirchherr J, Piscicelli L, Bour R, Kostense-Smit E, Muller J, Huibrechtse-Truijens A, Hekkert M (2018) Barriers to the circular economy: evidence from the European Union (EU). Ecol Econ 150:264–272. https://doi.org/10.1016/j.ecolecon.2018.04.028

Korhonen J, Honkasalo A, Seppälä J (2018a) Circular economy: the concept and its limitations. Ecol Econ 143:37–46. https://doi.org/10.1016/j.ecolecon.2017.06.041

Korhonen, J, Nuur, C, Feldmann, A, Birkie SE (2018b) Circular economy as an essentially contested concept. J Clean Prod 175:544–552.https://doi.org/10.1016/j.jclepro.2017.12.111

Lawrence MG, Williams S, Nanz P, Renn O (2022) Characteristics, potentials, and challenges of transdisciplinary research. One Earth 5(1):44–61. https://doi.org/10.1016/j.oneear.2021.12.010

Lazarevic D, Valve H (2017) Narrating expectations for the circular economy: towards a common and contested European transition. Energy Res Soc Sci 31:60–69. https://doi.org/10.1016/j.erss.2017.05.006

Leipold S, Weldner K, Hohl M (2021) Do we need a 'circular society'? Competing narratives of the circular economy in the French food sector. Ecol Econ 187:107086. https://doi.org/10.1016/j.ecolecon.2021.107086
Meadows DH (1999) Leverage points: places to intervene in a system. The Sustainability Institute: Hartland
Meadows DL, Meadows DH, Zahn E, Milling P (1972) Die Grenzen des Wachstums: Bericht des Club of Rome zur Lage der Menschheit. Stuttgart
Meadows DH, Randers J, Meadows DL (2004) Limits to growth: The 30-year update. White River Junction
Melles G (2021) Figuring the transition from circular economy to circular society in Australia. Sustainability 13(19):10601. https://doi.org/10.3390/su131910601
Millar N, McLaughlin E, Börger T (2019) The circular economy: swings and roundabouts? Ecol Econ 158:11–19. https://doi.org/10.1016/j.ecolecon.2018.12.012
Moreau V, Sahakian M, van Griethuysen P, Vuille F (2017) Coming full circle: why social and institutional dimensions matter for the circular economy. J Ind Ecol 21(3):497–506. https://doi.org/10.1111/jiec.12598
Murray A, Skene K, Haynes K (2017) The Circular Economy: an interdisciplinary exploration of the concept and application in a global context. J Bus Ethics 140(3):369–380. https://doi.org/10.1007/s10551-015-2693-2
Nelson HG, Stolterman E (2012) The design way. Intentional Change in an Unpredictable World, Cambridge/MA: MIT Press
Niskanen J, Anshelm J, McLaren D (2020) Local conflicts and national consensus: the strange case of circular economy in Sweden. J Clean Prod 261:121117. https://doi.org/10.1016/j.jclepro.2020.121117
Norman-Hansen J, Jensen BE, Jung M, Holm M (2021) Designing the irresistible circular society (White Paper).https://cdn-bloxhub20.pressidium.com/wp-content/uploads/2021/09/Creative-Denmark_white-paper_Designing-the-irresistible-circular-society.pdf
Pál V (2022) Social and cultural aspects of circular economy: towards solidarity and inclusivity. Routledge
Parrique T, Barth J, Briens F, Kerschner C, Kraus-Polk A, Kuokkanen A, Spangenberg JH (2019) Decoupling debunked. Evidence and arguments against green growth as a sole strategy for sustainability
Perry C, Jensen O (2001) Approaches to combining induction and deduction in one research study
Peukert D, Vilsmaier U (2019) Entwurfsbasierte Interventionen in der transdisziplinären Forschung. In Interventionsforschung. Springer, Wiesbaden, pp 227–250
Peukert D, Vilsmaier U (2021) Collaborative design prototyping in transdisciplinary research: an approach to heterogeneity and unknowns. Futures 132:102808. https://doi.org/10.1016/j.futures.2021.102808
Phaal R, Farrukh CJ, Probert DR (2004) Technology roadmapping—a planning framework for evolution and revolution. Technol Forecast Soc Chang 71(1–2):5–26
Prendeville S, Koria M (2022) Design discourses of transformation. She Ji: J Des, Econ, Innov 8(1):65–92. https://doi.org/10.1016/j.sheji.2022.01.002
Quist JN (2002) A strategic approach to radical sustainable innovations: stakeholder involvement, visioning, backcasting, learning and engineering education. In: Engineering education in sustainable development-proceedings. Delft University of Technology, pp 626–637
Quist J, Thissen W, Vergragt PJ (2011) The impact and spin-off of participatory backcasting: From vision to niche. Technol Forecast Soc Chang 78(5):883–897. https://doi.org/10.1016/j.techfore.2011.01.011
Raworth K (2017) Doughnut economics: seven ways to think like a 21st-century economist. Chelsea Green Publishing
Reike D, Vermeulen WJ, Witjes S (2018) The circular economy: new or refurbished as CE 3.0? Exploring controversies in the conceptualization of the circular economy through a focus on history and resource value retention options. Resour Conserv Recycl 135:246–264. https://doi.org/10.1016/j.resconrec.2017.08.027

Santa-Maria T, Vermeulen WJ, Baumgartner RJ (2022) The circular sprint: circular business model innovation through design thinking. J Clean Prod 362. https://doi.org/10.1016/j.jclepro.2022.132323

Schneider F, Tribaldos T, Adler C, Biggs R, de Bremond A, Buser T, Krug C, Loutre M, Moore S, Norstro A, Paulavets K, Urbach D, Spehn E, Wülse G, Zondervan R (2021) Co-production of knowledge and sustainability transformations: a strategic compass for global research networks. Curr Opin Environ Sustain 49:127–142. https://doi.org/10.1016/j.cosust.2021.04.007

Schröder P, Bengtsson M, Cohen M, Dewick P, Hofstetter J, Sarkis J (2019) Degrowth within–aligning circular economy and strong sustainability narratives. Resour Conserv Recycl 146:190–191. https://doi.org/10.1016/j.resconrec.2019.03.038

Schönfeld H (2020) Urban transformation design: Grundrisse einer zukunftsgewandten Raumpraxis. Basel: Birkhäuser

Seitz T (2020) Design thinking and the new spirit of capitalism. in design thinking and the new spirit of capitalism. Palgrave Pivot, Cham, pp 83–98

Simon HA (1969) The sciences of the artificial. MIT Press, Cambridge

Simonse L (2017) Design roadmapping. BIS Publishers

Steelman T, Bogdan A, Mantyka-Pringle C, Bradford L, Reed MG, Baines S, …, Poelzer G (2021) Evaluating transdisciplinary research practices: insights from social network analysis. Sustain Sci 16(2):631–645.https://doi.org/10.1007/s11625-020-00901-y

Temesgen A, Storsletten V, Jakobsen O (2019) Circular economy–reducing symptoms or radical change? Philos Manag, 1–20. https://doi.org/10.1007/s40926-019-00112-1

Temesgen A, Storsletten V, Jakobsen O (2021) Circular economy – reducing symptoms or radical change? Philos Manag 20(1):37–56. https://doi.org/10.1007/s40926-019-00112-1

Utrecht University (2020) Online Utrecht Degrowth Symposium: From circular economy to circular society [digital symposium].https://www.uu.nl/en/events/online-utrecht-degrowth-symposium-from-circular-economy-to-circular-society

United Nations Conference on Environment and Development (1992) Rio Declaration on Environment and Development. Sabah, Malaysia

Van der Bijl-Brouwer M, Malcolm B (2020) Systemic design principles in social innovation: a study of expert practices and design rationales. She Ji: J Des Econ Innov 6(3):386–407. https://doi.org/10.1016/j.sheji.2020.06.001

Vanhuyse F, Fejzić E, Ddiba D, Henrysson M (2021) The lack of social impact considerations in transitioning towards urban circular economies: a scoping review. Sustain Cities Soc 75:103394. https://doi.org/10.1016/j.scs.2021.103394

Velenturf AP, Purnell P (2021) Principles for a sustainable circular economy. Sustain Prod Consum 27:1437–1457. https://doi.org/10.1016/j.spc.2021.02.018

Vidal RVV (2004) The vision conference: facilitating creative processes. Syst Pract Action Res 17(5):385–405. https://doi.org/10.1007/s11213-004-5786-x

WBGU (2011) World in transition – a social contract for sustainability [flagship report]

WGBU (Hg.) (2020) Landwende im Anthropozän: Von der Konkurrenz zur Integration [flagship report]. Berlin

Wiek A, Iwaniec D (2014) Quality criteria for visions and visioning in sustainability science. Sustain Sci 9(4):497–512. https://doi.org/10.1007/s11625-013-0208-6

Zink T, Geyer R (2017) Circular economy rebound. J Ind Ecol 21(3):593–602. https://doi.org/10.1111/jiec.12545

Zwiers J, Jaeger-Erben M, Hofmann F (2020) Circular literacy. A knowledge-based approach to the circular economy. Cult Organ 26(2):121–141. https://doi.org/10.1080/14759551.2019.1709065

GPSR Compliance
The European Union's (EU) General Product Safety Regulation (GPSR) is a set of rules that requires consumer products to be safe and our obligations to ensure this.

If you have any concerns about our products, you can contact us on

ProductSafety@springernature.com

In case Publisher is established outside the EU, the EU authorized representative is:

Springer Nature Customer Service Center GmbH
Europaplatz 3
69115 Heidelberg, Germany

www.ingramcontent.com/pod-product-compliance
Ingram Content Group UK Ltd.
Pitfield, Milton Keynes, MK11 3LW, UK
UKHW022211051026
15624UKWH00001B/116

* 9 7 8 9 8 1 9 9 7 5 3 1 0 *